18th Annual Freight and Logistics Symposium

Thinking Globally, Acting Locally: Leveraging Minnesota's Strengths to Confront Future Freight and Logistics Challenges

Summary Report & Presentation

Minneapolis, Minnesota, USA
4 December 2015

Editors:

Michael McCarthy

ISBN: 978-1-5108-2179-8

Printed from e-media with permission by:

Curran Associates, Inc.
57 Morehouse Lane
Red Hook, NY 12571

Some format issues inherent in the e-media version may also appear in this print version.

Printed by Curran Associates, Inc. (2016)

For permission requests, please contact University of Minnesota Center for Transportation Studies at the address below.

University of Minnesota Center for Transportation Studies
200 Transportation & Safety Building
511 Washington Ave SE
Minneapolis, MN 55455
USA

Phone: 612-626-1077
Fax: 612-625-6381

cts@umn.edu

Additional copies of this publication are available from:

Curran Associates, Inc.
57 Morehouse Lane
Red Hook, NY 12571 USA
Phone: 845-758-0400
Fax: 845-758-2633
Email: curran@proceedings.com
Web: www.proceedings.com

TABLE OF CONTENTS

18th Annual Freight and Logistics Symposium

Thinking Globally, Acting Locally

Leveraging Minnesota's Strengths to Confront Future Freight and Logistics Challenges

A Summary Report | December 4, 2015 | Minneapolis, Minnesota

UNIVERSITY OF MINNESOTA

CENTER FOR TRANSPORTATION STUDIES

Sponsored by:
Center for Transportation Studies

Facilitated by:
Continuing Professional Education, University of Minnesota

In cooperation with:
Minnesota Department of Transportation

Minnesota Freight Advisory Committee

Council of Supply Chain Management Professionals–Twin Cities Roundtable

Metropolitan Council

Transportation Club of Minneapolis and St. Paul

18th Annual Freight and Logistics Symposium
Thinking Globally, Acting Locally

Leveraging Minnesota's Strengths to Confront Future Freight and Logistics Challenges

Introduction

Today's global economy is fast-moving and more competitive than ever, and it also is more metropolitan-centered than ever. The ability of the nation to prosper is increasingly dependent on the economic performance of its urban areas, which account for a large amount of the economic productivity in the United States today. But much of the critical infrastructure connecting U.S. metropolitan areas to resources and markets resides in rural regions. Federal, state, and local transportation programs must reflect the increasing interdependence between rural and urban places by maintaining and improving these vital linkages to ensure we can move goods efficiently and reliably.

Leaders and experts from all walks of transportation came together at the 18th Annual Freight and Logistics Symposium to discuss the challenges, progress to date, and future opportunities the industry faces as the United States shifts from a highway-centric transportation program to one that is becoming multimodal-focused in response to changing economic and demographic trends.

Presentations

Economic Competitiveness and Freight's Metropolitan Future

- Adie Tomer, Fellow, Metropolitan Policy Program, The Brookings Institution

MnDOT Update: Implementing the New Statewide Freight Plan and Action Agenda

- Bill Gardner, Director, Office of Freight and Commercial Vehicle Operations, MnDOT

Panel Discussions

Local Perspective on Global Economic Shifts

Moderator: Mark Berndt, Consultant and Outside Sales Representative, Quetica, LLC
Participants:

- Vanta E. Coda II, Executive Director, Duluth Seaway Port Authority
- Emily Jerve, Research Analyst, Minnesota Department of Agriculture
- Lorrie J. Louder, Senior Vice President of Business & Intergovernmental Affairs, Saint Paul Port Authority
- Adam Redlin, President, Midwest Global Trade Association

Implementing the New Statewide Freight Action Agenda

Moderator: Gina Baas, Associate Director, Center for Transportation Studies
Participants:

- Senator Scott Dibble, Minnesota Senate
- Bill Goins, Worldwide Account Manager, FedEx
- John Hauslauden, President, Minnesota Trucking Association
- Tim Henkel, Assistant Commissioner, Modal Planning and Program Management, MnDOT

Center for Transportation Studies
200 Transportation and Safety Building
511 Washington Avenue S.E., Minneapolis, MN 55455-0375
Phone: 612-626-1077 E-mail: cts@umn.edu Web: cts.umn.edu

Editor: Michael McCarthy **Writer:** Nancy Strege **Designer:** Angie Kronebusch
Photos: Shutterstock

Economic Competitiveness and Freight's Metropolitan Future

Domestic trade in the United States is vital to the health and vibrancy of the nation's economy, yet surprisingly, very little is known about this economic driver. With exports and imports of more than $3 trillion in international goods and a staggering $17 trillion worth of domestic goods traded between regions, these numbers show just how critical trade is to the U.S. overall, but in particular, the importance of the trade occurring within the country's borders.

National trade statistics generally look only at how the U.S. trades with other countries, explains Adie Tomer, a fellow at The Brookings Institution Metropolitan Policy Program and a member of the Metropolitan Infrastructure Initiative. "When an airplane built in Seattle by Boeing is sold to Delta Airlines in Atlanta ... we can't see that anywhere in our trade statistics. We think that's a major shortcoming in this data." Without a full understanding of how these trade networks work together, he adds, leaders in both the public and private sectors cannot develop targeted freight strategies and transportation investments to support the extensive supply chains that underpin the U.S. economy.

> Large metropolitan areas are where trade, particularly goods and services in advanced industries, concentrates in the United States.
> — Adie Tomer

Metro-level trade analysis

In its Metro Freight Series, Brookings Institution researchers, including Tomer, addressed these deficiencies in their study of various aspects of trading relationships, specifically, trade at the metropolitan level. Their findings help better explain how communities relate to each other and what their freight needs might be moving forward, and help drive home the importance of trade in economic terms. "We now have a powerful new way to look at trade data that tells us something about freight we didn't know before," Tomer says.

In particular, this research shows that large metropolitan areas are where trade, particularly goods and services in advanced industries, concentrates in the United States, with 80 percent of traded value starting or ending in these areas. "These are the power-packed locations that drive where we are going in the future—everything from industrial output, where people work, and where their families will grow up," Tomer says. "This has huge implications for how freight moves in our country and how we approach it."

More information

The Brookings Institution Metro Freight Series: **brookings.edu/research/reports2/2015/06/metro-freight**

Far-reaching costs of congestion

Because large cities are where the bulk of economic activity resides, it is not surprising that they also are where most traffic congestion occurs. And while clogged roadways

Thinking of freight at a metropolitan scale

The largest 100 metropolitan areas in the country take up only 12 percent of our land mass, but contain 66 percent of our population and represent three-quarters of our GDP.

inconvenience and frustrate commuters, they also impede the movement of trillions of dollars in goods traveling between different markets. In fact, Tomer points out, the congestion in one market can actually hurt regions hundreds or even thousands of miles away.

And, the problem will only get worse. The Brookings research indicates that urban trucking is continuing to grow, and already is far outpacing passenger vehicle-miles traveled. This increase directly coincides with the rise of e-commerce and the use of digital communications to manage shipping for logistics firms such as UPS and FedEx and major private shippers such as Walmart.

Metropolitan freight planning

Historically, Washington has not paid attention to the areas of congestion around major freight hubs at urban cores, Tomer says, but it will be more important than ever for freight investment to be targeted at these choke points. Additionally, metropolitan leaders, along with their state and federal peers, must begin to "design policies to address the new reality of the modern metropolis."

While there is no blueprint for metropolitan freight planning across the country as of yet, Tomer concludes, the new Metro Freight research data available will greatly help direct these efforts. "With this data, combined with the U.S. Department of Transportation's new freight strategy—which I believe is the best freight document to come out of [Washington] D.C. in a while—we might finally get smart about freight investments."

> "The precision instrument industry is your economic lifeblood in many ways, even though in some ways it is a relatively small industry."
> — Adie Tomer

Brookings analysis illustrates vital role for freight in Minnesota and Minneapolis–St. Paul metro

Adie Tomer, citing new data from the Brookings Metro Freight Series, says Minneapolis–St. Paul is the primary freight hub in the Upper Midwest. The metro area produces and consumes a wide range of commodities, with an extensive freight infrastructure that transports goods to and from regional markets and around the world.

The total goods trade for Minnesota involves $393 billion worth of goods moving in and out of the state. While it's an average freight-producing state with several clear industrial strengths, there is some disjointedness across industries. "Basically, Minnesota has certain things it does well and certain things it does not do so well," Tomer explains.

Tomer illustrates this using industry location quotients (LQs), which are one way of quantifying how concentrated an industry is in a region compared to a larger geographic area, such as the state or nation. Industries with high LQs and relatively high total job numbers typically form a region's economic base. For example, the LQ for energy products in Minnesota is 0.3, indicating that the state mostly buys petroleum and coal from other parts of the country and the world. Similarly, the LQ for transportation equipment is 0.4, which suggests that Minnesota buys this equipment from other places and does not manufacture most of it.

In contrast, the LQ for precision instruments in Minnesota is 2.5. "That is off the charts," Tomer says. "This shows that [the precision instrument] industry is your economic lifeblood in many ways, even though in some ways it is a relatively small industry." Other Minnesota industries with relatively high LQs include agriculture, food, and wood industries. "These will continue to produce great value for the state as well."

Perspectives on Keeping Pace with the World Market

Minnesota was once well-known for its raw-material-based economy dominated for more than a century by the mining of most of the world's soft-iron ore. The state's economy has since evolved to one based on finished, tradable goods and services that have helped Minnesota remain a viable player in the global marketplace. Area experts describe some of the challenges and opportunities confronting the state as it works to keep pace and succeed in the world market.

The power of ag

Agriculture is one key force behind Minnesota's vibrant economy, with about a third of the state's cash receipts generated from agricultural exports, says Emily Jerve, research analyst with the Minnesota Department of Agriculture. The department works directly with various businesses, including food processors, packagers, and producers, to help move these important assets to market.

The Ag Department also embraces a number of partnerships with state and regional trade groups and other federal and state agencies. "We tap into these when planning trade or buyers' missions," Jerve explains. "But the broader point is that we have a lot of networks—this is where we get the most value for Minnesota companies to help promote agriculture exports."

Many of these ag exports move by barge down the Mississippi River, starting at the St. Paul Harbor. The much-anticipated opening of a third Panama Canal channel sometime this year will open up major Asian markets to Minnesota shippers, adds Lorrie Louder of the Saint Paul Port Authority. For a port city like St. Paul, which is more than 1,300 miles inland, "the time and cost savings will be enormous."

Infrastructure demands

While the Panama Canal expansion is a positive, the big challenge of keeping the infrastructure firm and healthy remains. "Our business customers tell us that if we can't provide [reliable infrastructure], in this case, dock walls, then they will take their business elsewhere," Louder says.

Every four years, the American Society of Civil Engineers comes out with a report card depicting the condition and performance of the transportation infrastructure in the United States. "We are at a D grade nationally—Minnesota is at a C," says Vanta Coda, executive director of the Duluth Seaway Port Authority. "State by state we are going it alone. We have to talk about an integrated policy, but we're fragmented, and that's not going to work. Ultimately, to get back to number one, we need to lift together."

Consistent freight flows

Along with the aging infrastructure problem, first- and last-mile connectors—including roads, railways, bridges, and tunnels—are falling behind 21st century needs and are threatening state and national global competitiveness. Adam Redlin, president of the Midwest Global Trade Association, stresses the importance of maintaining consistent freight flows and reducing infrastructure failures. "Every single time we have an outage," he says, "we open the door to increased competition from other countries."

Agricultural goods, in particular, are very time sensitive. "This is why, although we don't directly play a role in the transportation planning, as far as an overall well-defined network approach to freight investment in infrastructure in the U.S., we support some type of long-term strategy," Jerve says.

Critical public/private partnerships

"Supposedly, we have an intermodal transportation system, but the money and planning does not flow that way," adds Mark Berndt, with Quetica, LLC. "We are moving in that direction, but we are not there yet. We have to advance the notion of the public sector funding more last-mile [infrastructure] and use those public/private partnerships to develop modes across systems."

Louder, who agrees that such public/private partnerships are critically important, predicts new and unique collaborations will shape the future for major freight investments across the country. "Since the [federal government] is not as engaged as we would like it to be, and we know there are challenges with an aging infrastructure," she says, "we have to pull ourselves together and work to create some consistency in our partnerships."

“We need to find ways to make the freight transportation system more efficient.”
— Bill Gardner

MnDOT Shares New Statewide Freight Plan and Action Agenda

Between now and 2040, annual freight tonnage in Minnesota is expected to grow by 80 percent and place exponentially greater stress on an already-congested system. Projections like this make it critical that the Minnesota Department of Transportation (MnDOT) plans now to maintain an integrated, efficient, multimodal transportation system in the future. For these reasons, Minnesota recently updated its statewide freight system plan, which provides a new policy framework and strategies to guide planning and investment in various transportation modes.

To explain the plan, Bill Gardner, director of the MnDOT Office of Freight and Commercial Vehicle Operations, points to the newly developed freight action agenda, a document that houses all of the freight plan recommendations in a single place. Moving forward, the action agenda will serve as a guide for plan implementation and will drive freight system improvements for the next 5 to 10 years.

Reducing transportation costs

As part of the freight plan update, MnDOT gathered input and gained insights from a variety of stakeholders via outreach efforts including interviews and case studies with key players in the freight community. “One of the benefits of doing a plan like this is to energize our stakeholders and broaden the circle of people we normally talk to,” Gardner says. “What we heard consistently throughout [our discussions with these groups] is that we need to find ways to make the freight transportation system more efficient, specifically to reduce the cost of transportation for Minnesota shippers.”

Minnesota shippers’ main concerns revolve around system reliability, safety, congestion, bridge and pavement condition, and economic development, Gardner reports. “There were needs identified for all modes, but highway issues topped the list,” he says. “We’ve done a good job reaching consensus regarding some of the issues we face as a state for freight, and also the strategies that are identified in the freight plan as ones that we can rally around.”

Identifying key strategies

The freight action agenda identifies 30 strategies for advancing freight in Minnesota, with overarching objectives related to partnerships, communication, system connectivity, asset management, safety, and freight and community balance. With 30 strategies, this plan is somewhat aspirational, Gardner says, and advancing the policies within it will require a lot of different stakeholders to come together.

“As a DOT, we tend to default to discussion about infrastructure and funding that infrastructure, which is an important discussion,” he adds. “But what I have heard today is that there are a whole variety of public policy issues—be they workforce development, energy policy, industrial development policy—that are part and parcel to the whole transportation system.”

While such issues are not the bread and butter of the DOT, Gardner concludes, they are areas MnDOT is willing to work on with others.

Minnesota’s family of plans

In addition to the statewide freight plan, MnDOT is in the process of updating the Statewide Multimodal Transportation Plan and the Minnesota State Highway Investment Plan. All of these plans are components of the Minnesota GO 50-Year Vision for Transportation, which focuses on quality of life, environmental health, and economic competitiveness.

Minnesota GO 50-year vision: **dot.state.mn.us/minnesotago/vision.html**

State Freight System Plan: **dot.state.mn.us/planning/freightplan**

Statewide Multimodal Transportation Plan (SMTP) and the Minnesota State Highway Investment (MnSHIP): **dot.state.mn.us/planning/smtp-mnship**

Implementing the New Statewide Freight Action Agenda

We're all aware of our rapidly aging transportation infrastructure and its detrimental effect on freight movement in Minnesota and the nation. According to MnDOT assistant commissioner Tim Henkel, we need to have serious discussions about revenue in relation to freight and passenger travel in this state to implement the new statewide freight action agenda. Henkel, who oversees modal planning and program management, says without that conversation, MnDOT can only preserve the system, unable to tackle the many strategies in the action agenda.

"While I welcome the conversation about new [revenue] sources and other ways of doing business and I support efforts for us to look at those things that do not necessarily require investment," he says, "we need to think about the future needs of our transportation system and support additional funding to be able to meet those needs."

Need for public dialogue

Minnesota state Sen. Scott Dibble, chair of the Senate Transportation and Public Safety Committee, agrees, adding that there is particular need for more discussion among policymakers about the industries and commodities that currently drive the state's economy as well as new, emerging industries. Such a dialogue, he says, is necessary for government leaders to respond with tangible transportation-related policy changes and projects.

"Freight is an issue that legislators don't really think about a whole lot, except maybe in the abstract," Dibble says. "We talk about the need for improving roads and bridges, which is tied to the economy, but we really talk about it from how big and robust our roads will be, what is their capacity, what are their safety issues, what is the pavement condition, that sort of thing. I am grateful for the development of [the freight plan and action agenda] and am especially pleased to see these 30 action items that have come from it to give guidance to people like me."

Partnerships key to plan

Bill Goins, worldwide account manager with FedEx, reiterates the key role of partnerships in delivering on the strategies laid out in the freight plan and action agenda. "We all recognize the development of a freight plan is not a silo. A freight plan is really part of the integration of working with the transportation needs of commuters and others," Goins says. "Every car we can take off of our confined highway system helps us move goods in our state and helps us to remain competitive. It's important that we create an environment where people can get to jobs and entertainment and at the same time create a system that allows for that movement of goods."

One strategy identified in the freight action agenda that is of great interest to the trucking industry involves workforce development, adds John Hausladen, president of the Minnesota Trucking Association. "It is vital to include workforce issues in a freight plan," he says, "because right now, the American Trucking Association says we have a shortage of about 30,000 to 35,000 truck drivers nationally, which could grow to 240,000 by the year 2022."

Workforce development

Though there are a number of mitigating factors behind the driver shortage, one prevailing problem is the industry's over-reliance on an older cohort of experienced drivers who are moving through to retirement, Hausladen explains. "There are not enough younger drivers coming on board, partially because there is no real path from high school to over-the-road truck driving."

But Hausladen cites a provision in the FAST Act for a pilot program to help younger military veterans under 21 drive interstate commerce trucks as a "creative solution to bridge the gap and get younger people trained, in a safe environment." He also applaudes the inclusion of an action agenda item that deals with first- and last-mile issues, which generally are the most expensive, least efficient, and most problematic in terms of moving freight.

"I think the biggest challenge we have with infrastructure funding today is local government road funding as it relates to freight," concludes Hausladen, who serves with Henkel and Goins on the Minnesota Statewide Freight System Plan Advisory Committee. "I'm pleased that the freight plan addresses this."

national shortage of up to

35,000

truck drivers

Focus on the FAST Act

In December 2015, President Obama signed into law the Fixing America's Surface Transportation Act (FAST Act). The legislation authorizes $305 billion for federal surface transportation programs during fiscal years 2016 through 2020. Specifically, the FAST Act establishes both formula and discretionary grant programs to fund critical transportation projects designed to benefit freight movements. Of the $10 billion formula program, Minnesota will get between $150 and $200 million over five years.

"Even if you spend this on one project, this money won't go very far," explains Adie Tomer, with The Brookings Institution. "It is real money, don't get me wrong. But it's not going to change the state's freight plan."

MnDOT's Bill Gardner, however, views the funding as a way to strategically leverage other types of public and private resources for transportation. "Finally, there is an acknowledgement of a federal responsibility for the freight network," he says. "There is dedicated freight funding that I think really is a game changer. It's not a lot of money in terms of overall needs in the state, but it is a starting point."

Vanta Coda, with the Duluth Seaway Port Authority, expresses similar optimism. "This is a huge step forward. It doesn't fully fund everything, but there are real dollars for the first time," he points out, "and real discussion about creating maritime freight plans that are integrated with highway and rail."

> "This is a huge step forward. It doesn't fully fund everything, but there are real dollars for the first time and real discussion about creating maritime freight plans that are integrated with highway and rail."
> — Vanta Coda

Gardner, too, notes the policy and planning aspects of the FAST Act. "One of the big benefits of this [legislation] is that it is multimodal," he says. "Up until now, everything has been highly skewed to be highway-centric. It's a pretty complicated program, but overall, most people say this is a very good start."

The legislation also includes reauthorization of the U.S. Export-Import Bank of the United States (EXIM), which expired in June. EXIM finances and insures foreign purchases of U.S. goods for customers unable or unwilling to accept the credit risk. "Not many people understand the importance of this," Coda explains. "There was $27 billion worth of trade at risk."

In the end, Tomer acknowledges the progress in the FAST Act, but cautions that much remains to be done to improve U.S. freight planning and investment. "States know that the federal government is not coming to save them anytime soon," he concludes, "so they need to come up with money on their own."

More information about the FAST Act: **transportation.house.gov/fast-act**.

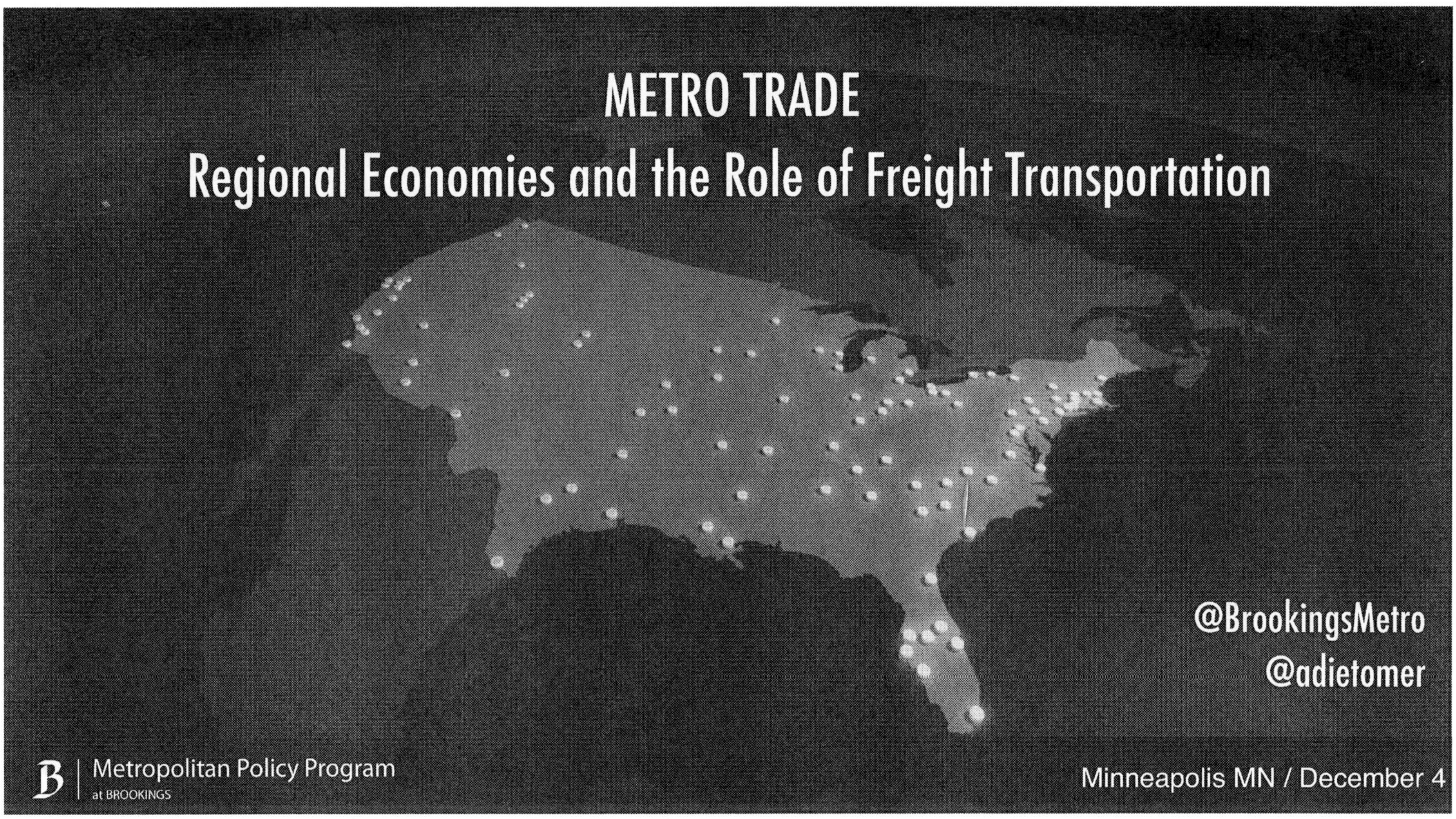
METRO TRADE
Regional Economies and the Role of Freight Transportation
@BrookingsMetro
@adietomer
Metropolitan Policy Program
at BROOKINGS
Minneapolis MN / December 4

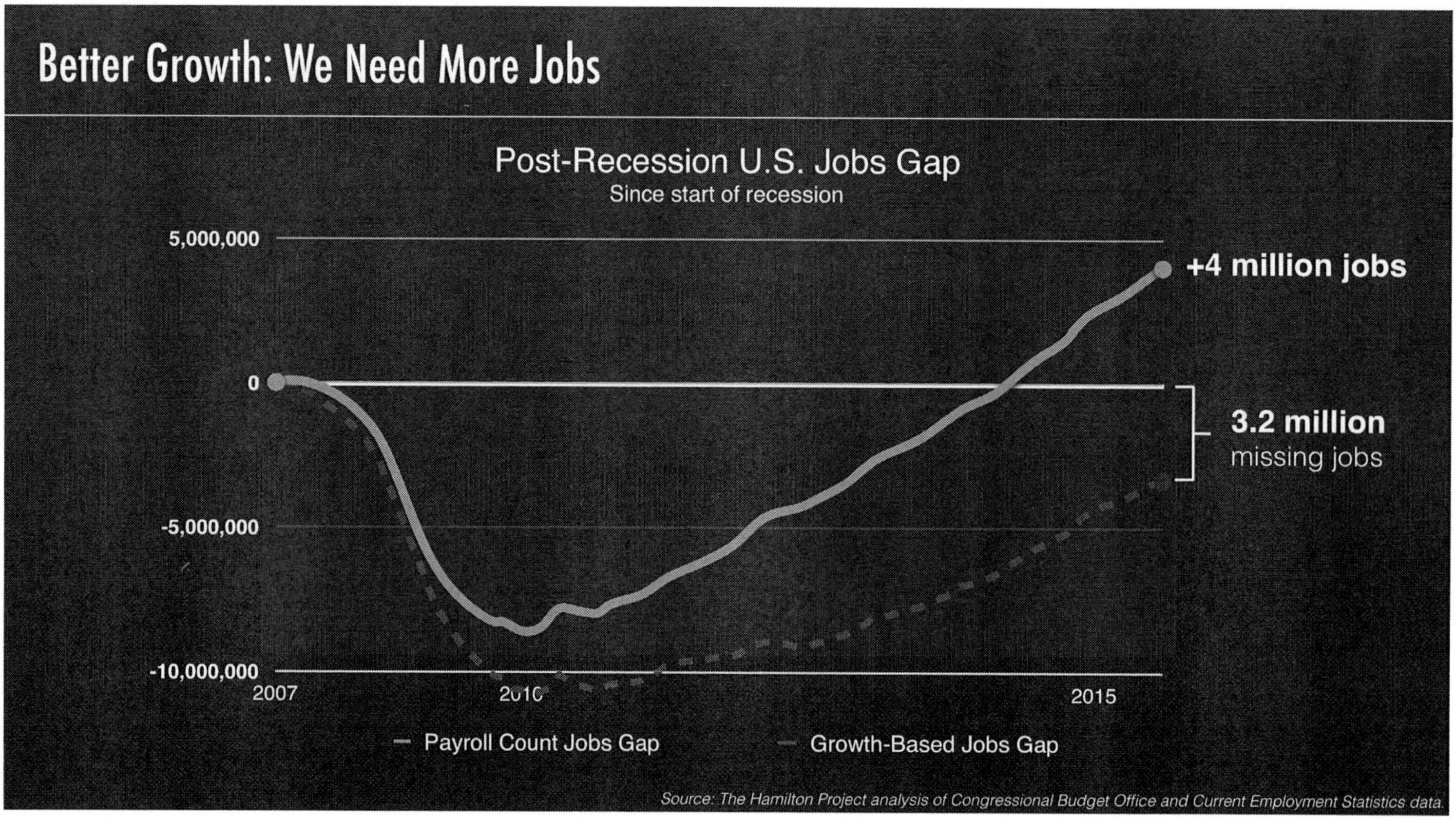
Better Growth: We Need More Jobs
Post-Recession U.S. Jobs Gap
Since start of recession
5,000,000
0
-5,000,000
-10,000,000
2007
2010
2015
+4 million jobs
3.2 million
missing jobs
Payroll Count Jobs Gap
Growth-Based Jobs Gap
Source: The Hamilton Project analysis of Congressional Budget Office and Current Employment Statistics data.

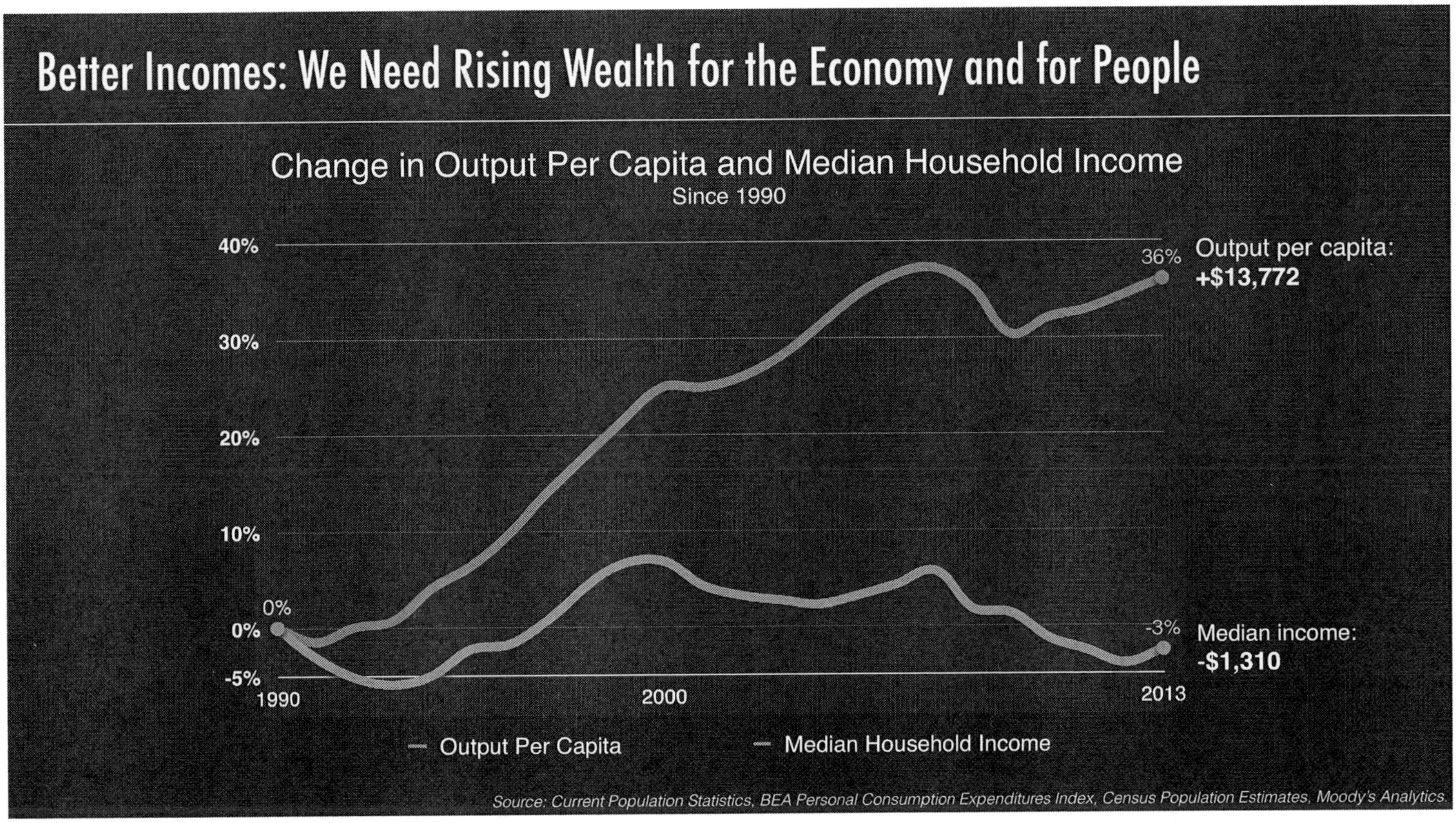
Better Incomes: We Need Rising Wealth for the Economy and for People
Change in Output Per Capita and Median Household Income
Since 1990
40%
30%
20%
10%
0%
-5%
1990
2000
2013
0%
36%
-3%
Output per capita:
+$13,772
Median income:
-$1,310
Output Per Capita
Median Household Income
Source: Current Population Statistics, BEA Personal Consumption Expenditures Index, Census Population Estimates, Moody's Analytics.

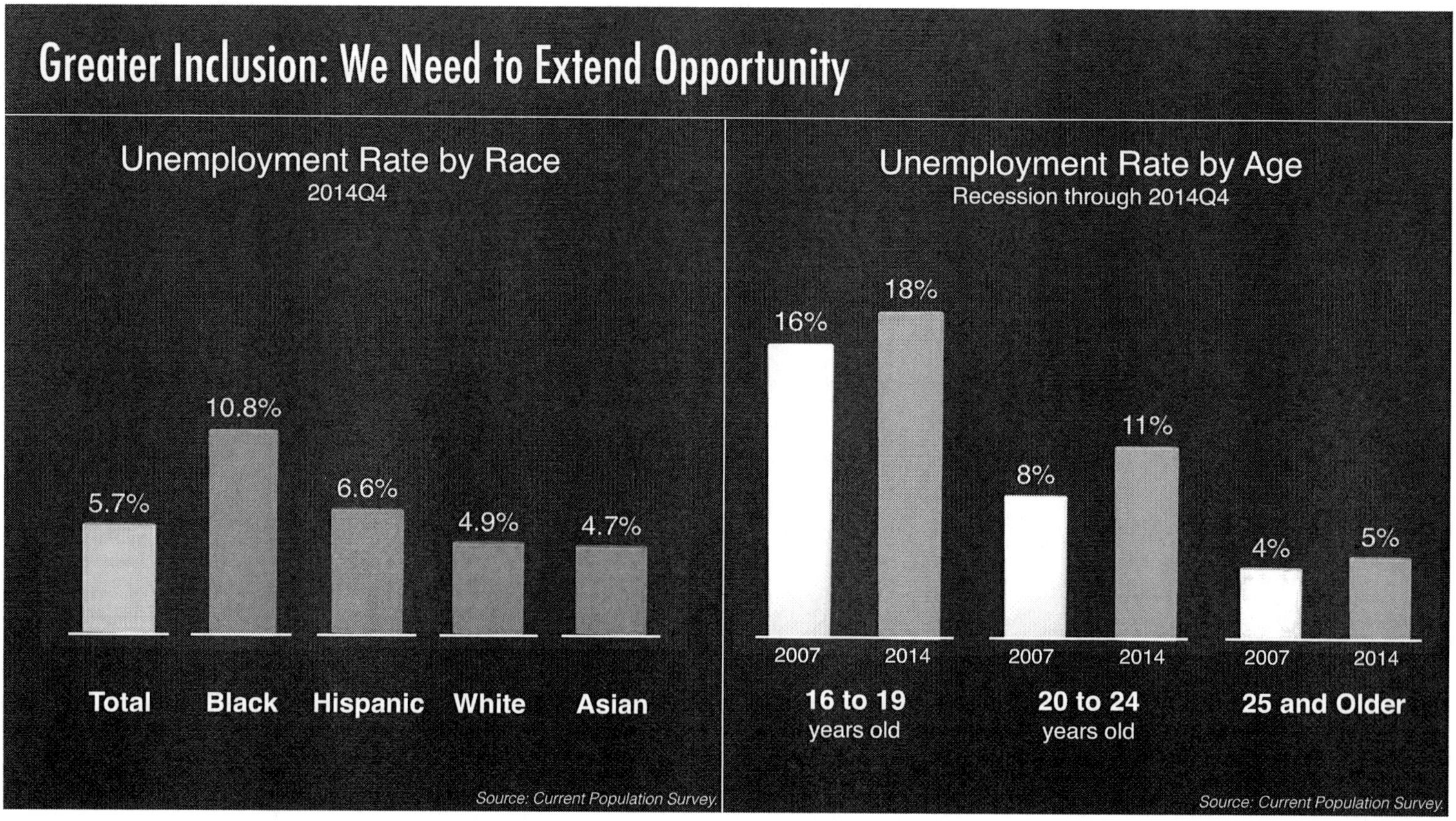
Greater Inclusion: We Need to Extend Opportunity
Unemployment Rate by Race
2014Q4
5.7%
10.8%
6.6%
4.9%
4.7%
Total
Black
Hispanic
White
Asian
Source: Current Population Survey.
Unemployment Rate by Age
Recession through 2014Q4
16%
18%
8%
11%
4%
5%
2007
2014
2007
2014
2007
2014
16 to 19
years old
20 to 24
years old
25 and Older
Source: Current Population Survey.

2 3

1 To prosper, public leaders are beginning to recognize the importance of trade

1 3

2 Trade prioritization requires a new perspective on freight trends

1 2

3 What's the future of American freight planning and investment?

2 3

1 To prosper, public leaders are beginning to recognize the importance of trade

Global Competition Is Fierce, with Growth Shifting Away from the United States
Global Economic Growth
22.8%
US
17.5%
BIC Countries
2013
Source: International Monetary Fund, World Economic Outlook Database, October 2013

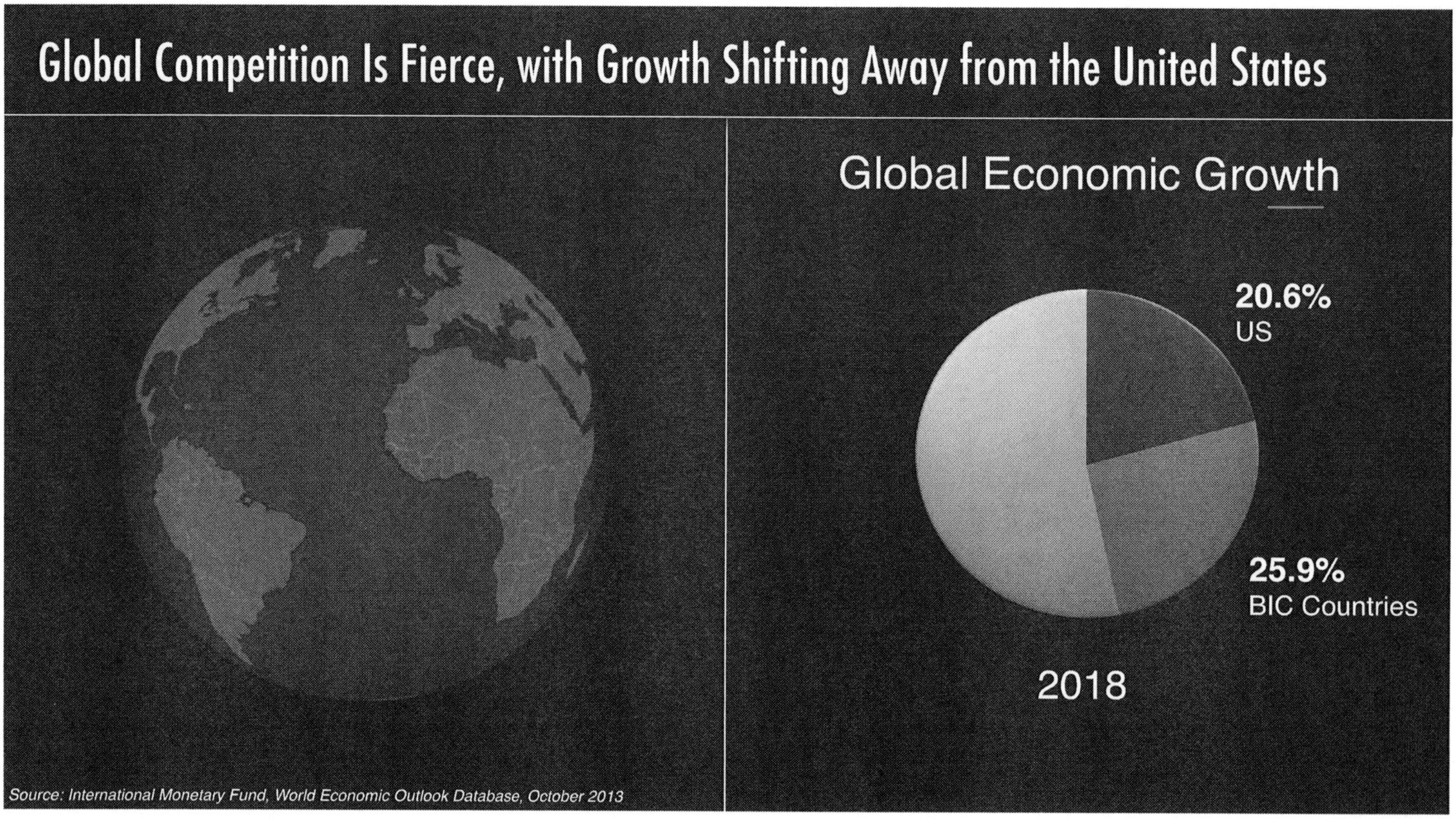
Global Competition Is Fierce, with Growth Shifting Away from the United States
Global Economic Growth
20.6%
US
25.9%
BIC Countries
2018
Source: International Monetary Fund, World Economic Outlook Database, October 2013

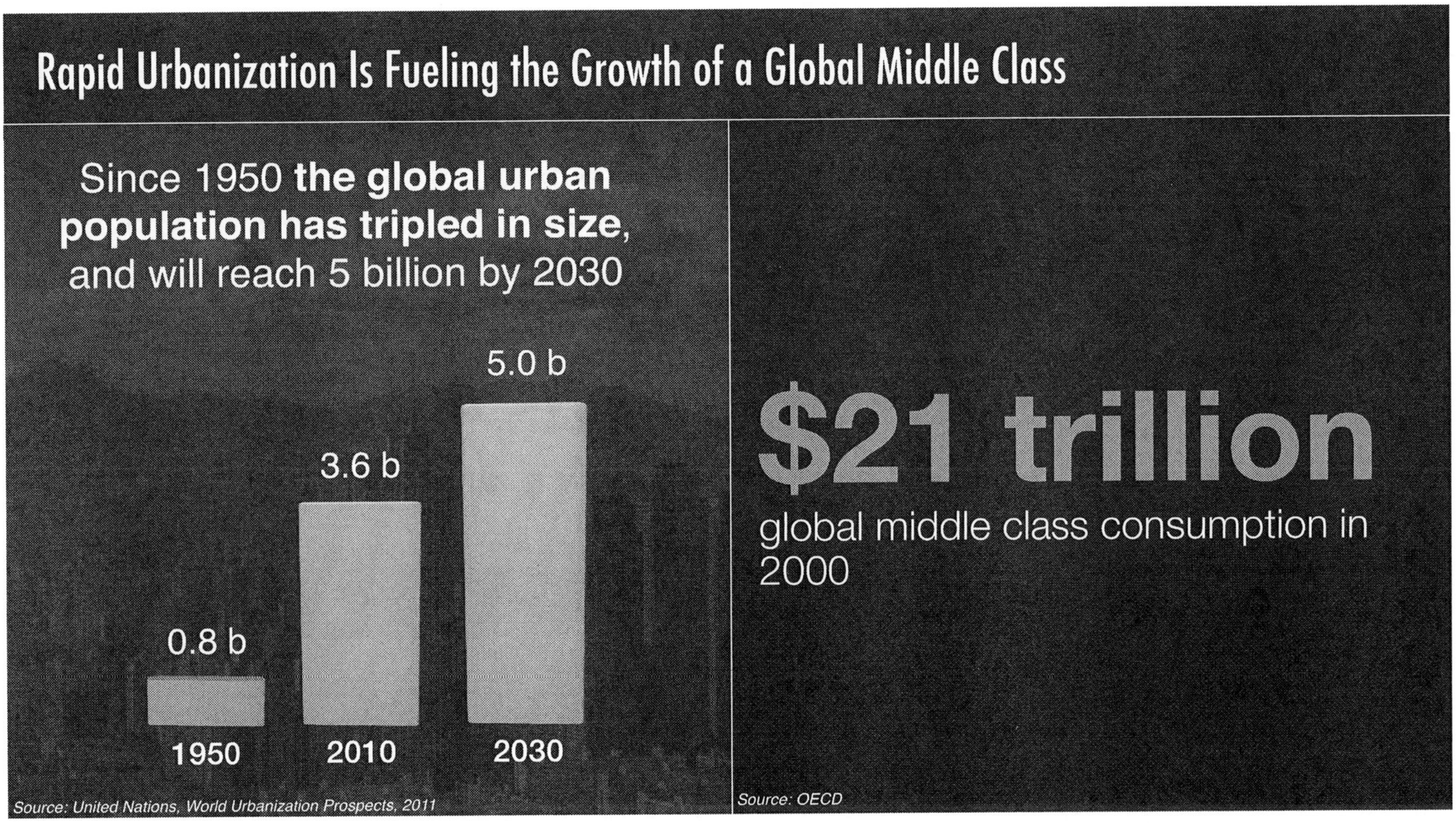
Rapid Urbanization Is Fueling the Growth of a Global Middle Class
Since 1950 the global urban population has tripled in size, and will reach 5 billion by 2030
5.0 b
3.6 b
0.8 b
1950
2010
2030
Source: United Nations, World Urbanization Prospects, 2011
$21 trillion
global middle class consumption in 2000
Source: OECD

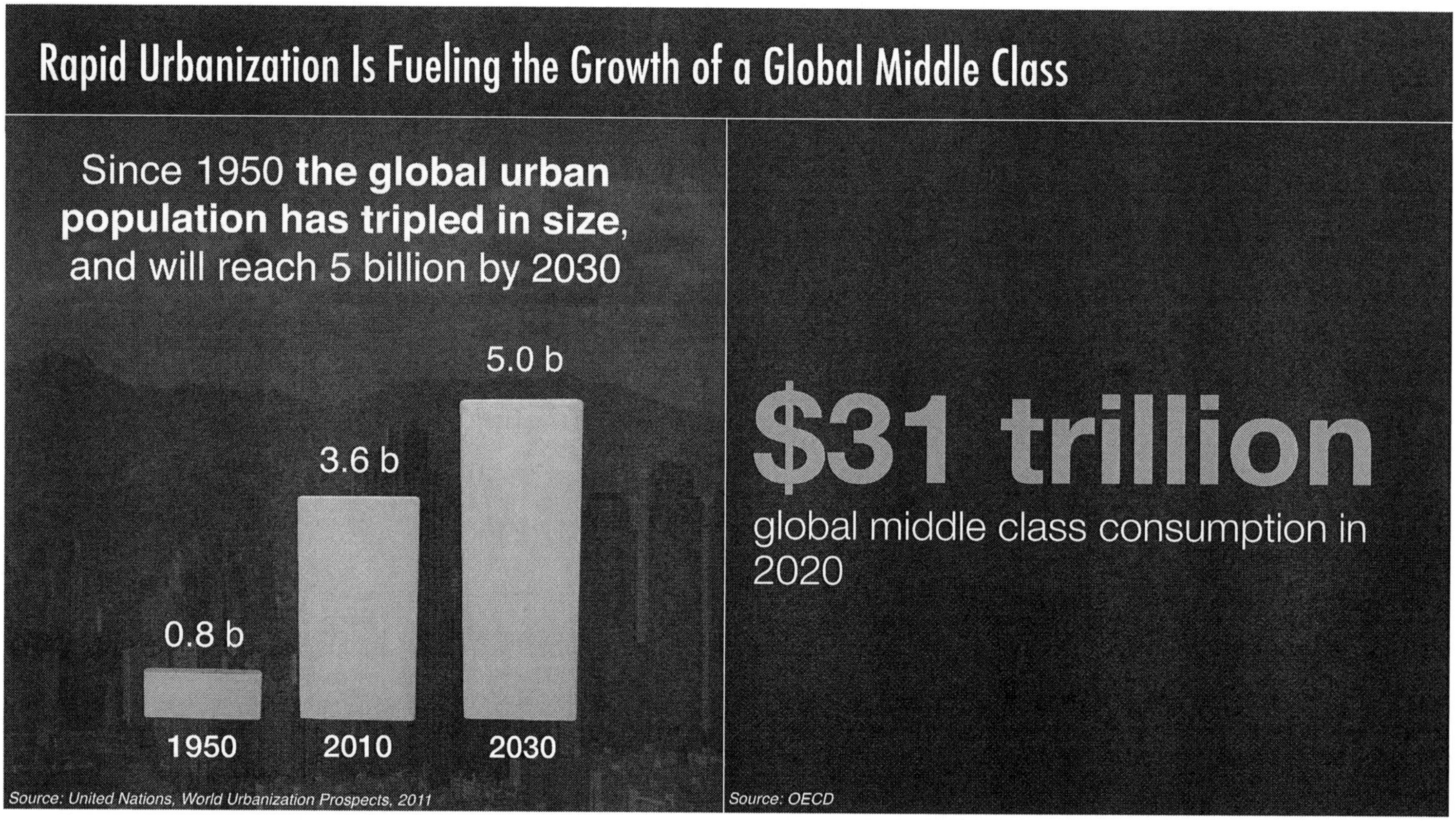
Rapid Urbanization Is Fueling the Growth of a Global Middle Class
Since 1950 the global urban population has tripled in size, and will reach 5 billion by 2030
5.0 b
3.6 b
0.8 b
1950
2010
2030
Source: United Nations, World Urbanization Prospects, 2011
$31 trillion
global middle class consumption in 2020
Source: OECD

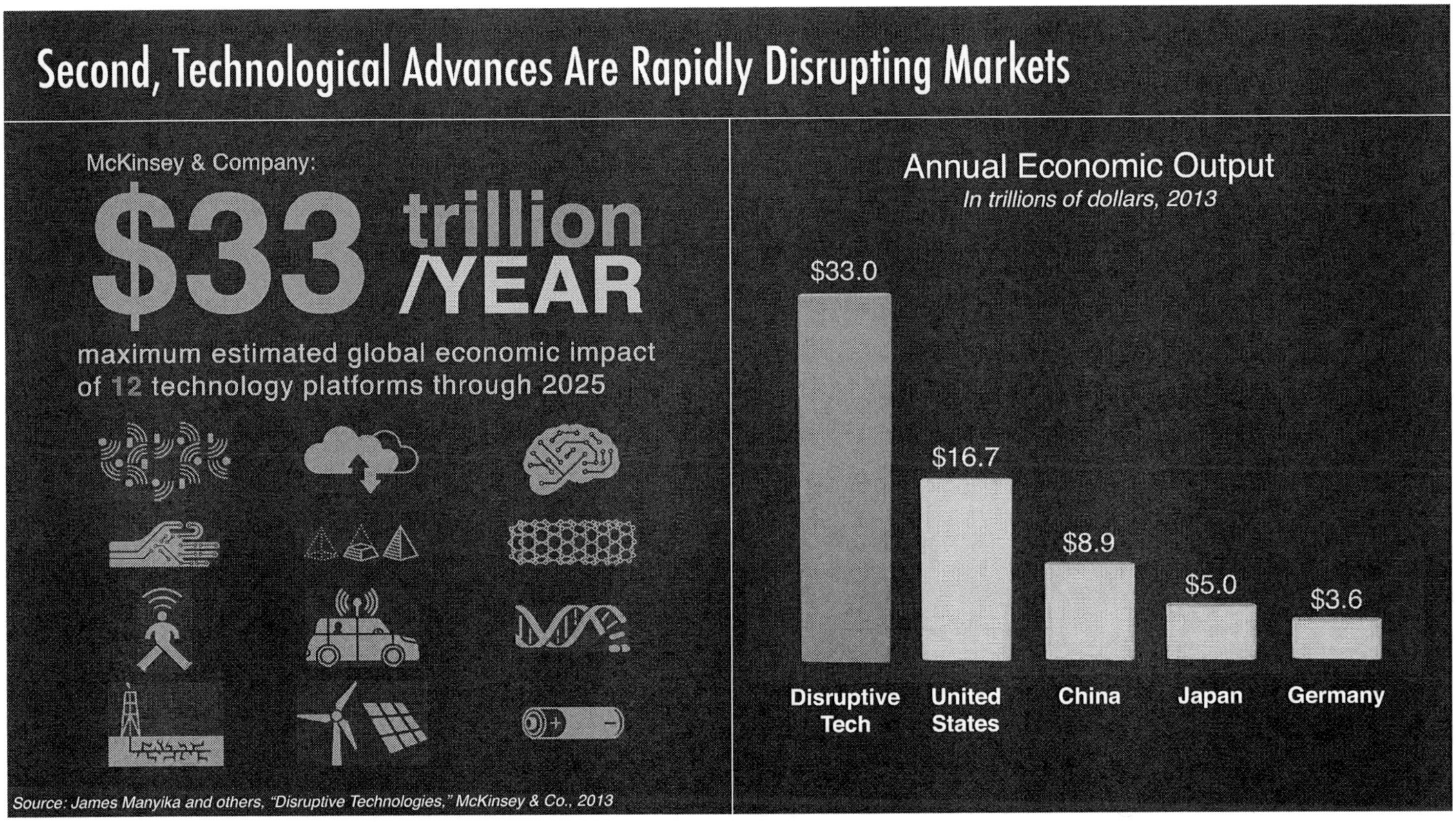
Second, Technological Advances Are Rapidly Disrupting Markets
McKinsey & Company:
$33 trillion /YEAR
maximum estimated global economic impact of 12 technology platforms through 2025
Source: James Manyika and others, "Disruptive Technologies," McKinsey & Co., 2013
Annual Economic Output
In trillions of dollars, 2013
$33.0
$16.7
$8.9
$5.0
$3.6
Disruptive Tech
United States
China
Japan
Germany

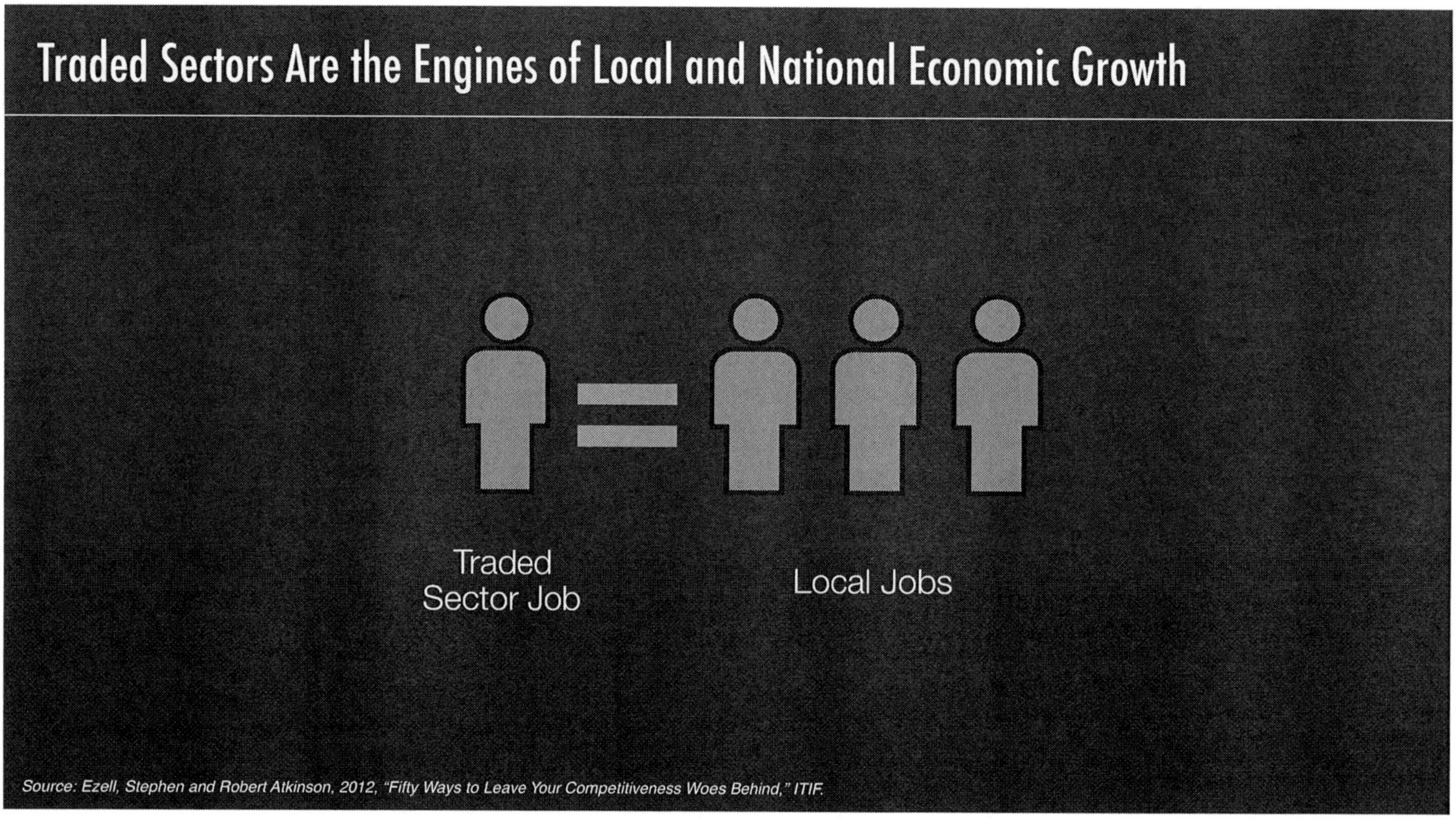
Traded Sectors Are the Engines of Local and National Economic Growth
Traded Sector Job
Local Jobs
Source: Ezell, Stephen and Robert Atkinson, 2012, "Fifty Ways to Leave Your Competitiveness Woes Behind," ITIF.

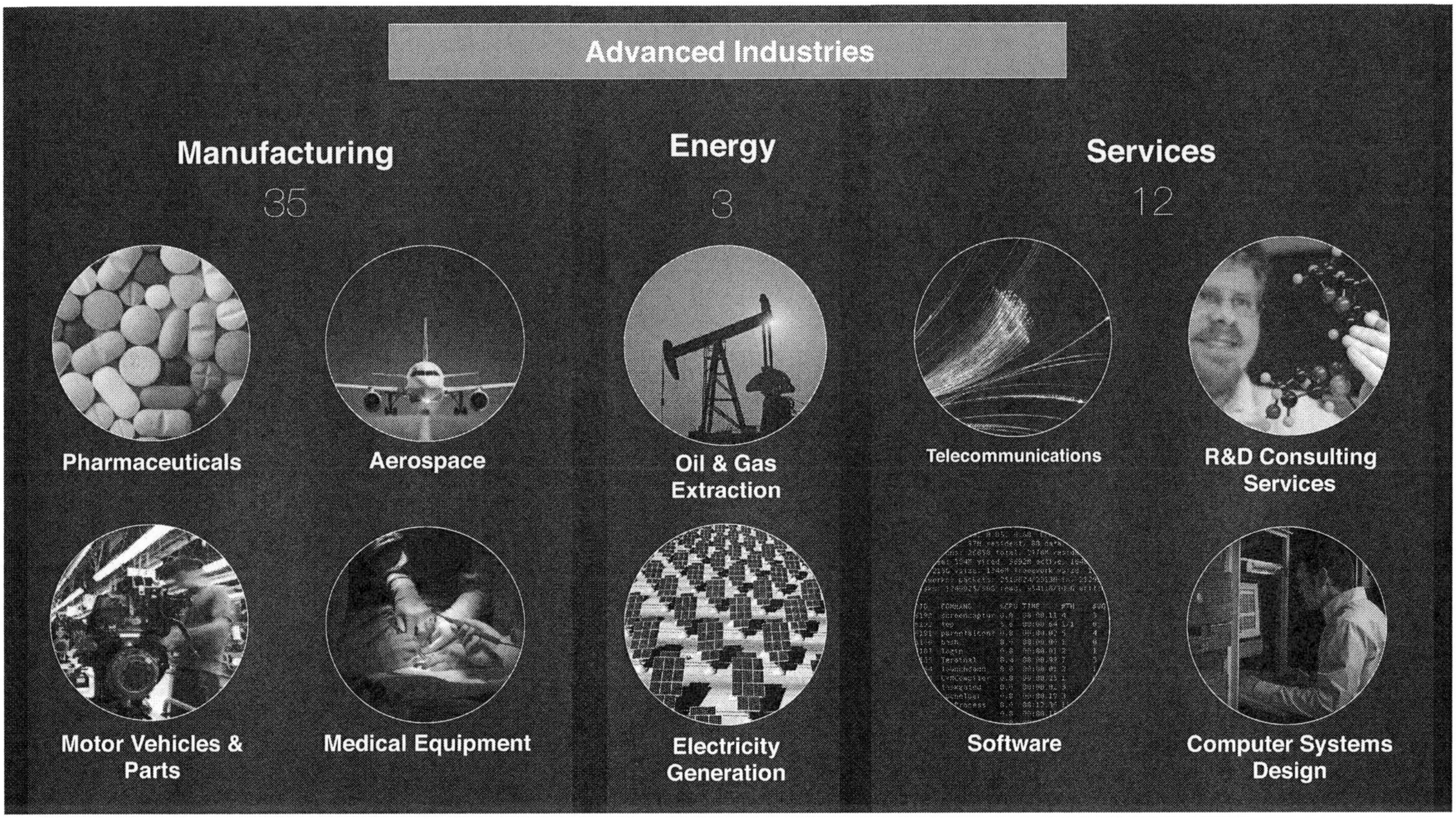
Advanced Industries
Manufacturing
35
Energy
3
Services
12
Pharmaceuticals
Aerospace
Oil & Gas Extraction
Telecommunications
R&D Consulting Services
Motor Vehicles & Parts
Medical Equipment
Electricity Generation
Software
Computer Systems Design

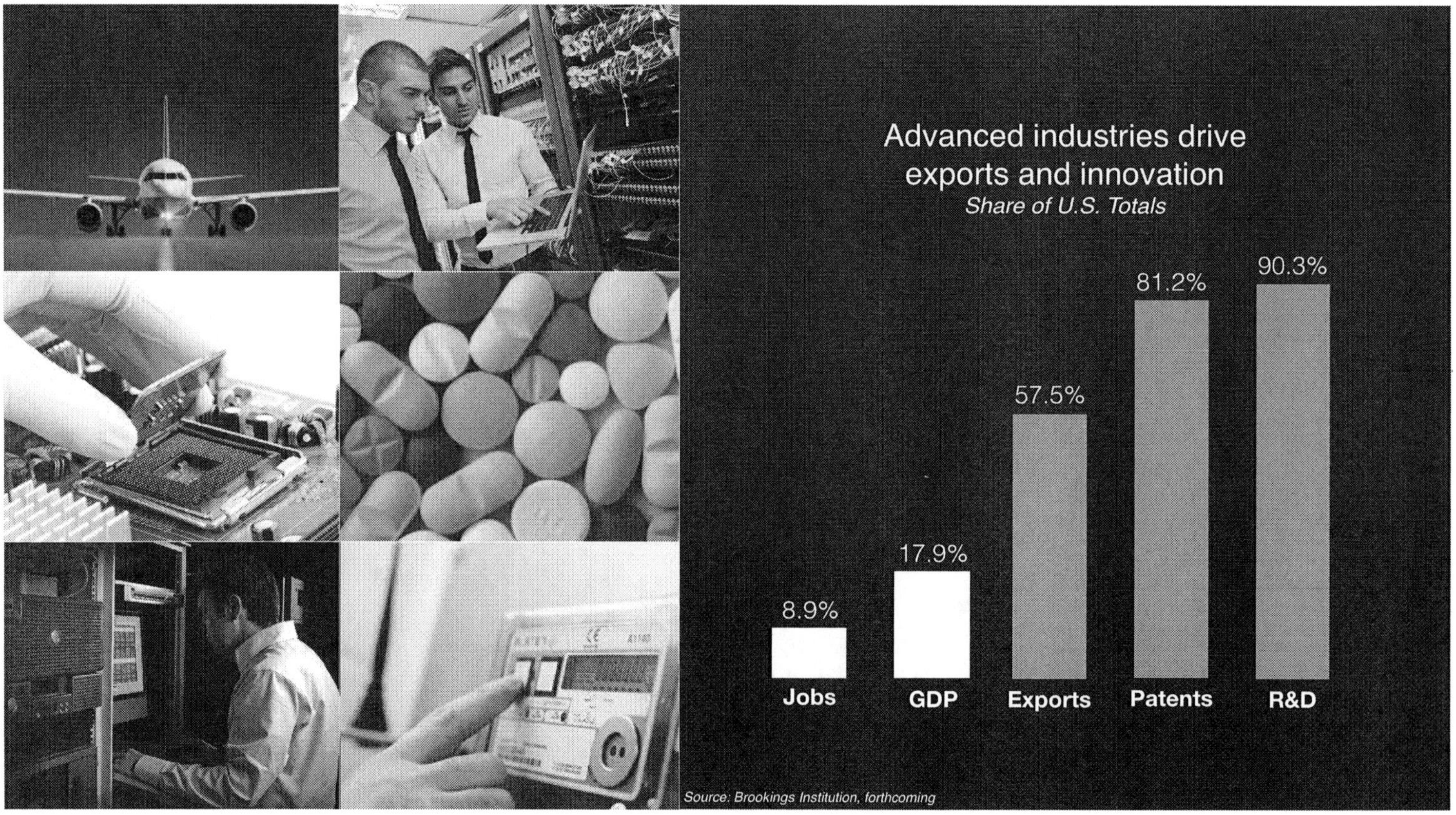
Advanced industries drive
exports and innovation
Share of U.S. Totals
8.9%
17.9%
57.5%
81.2%
90.3%
Jobs
GDP
Exports
Patents
R&D
Source: Brookings Institution, forthcoming

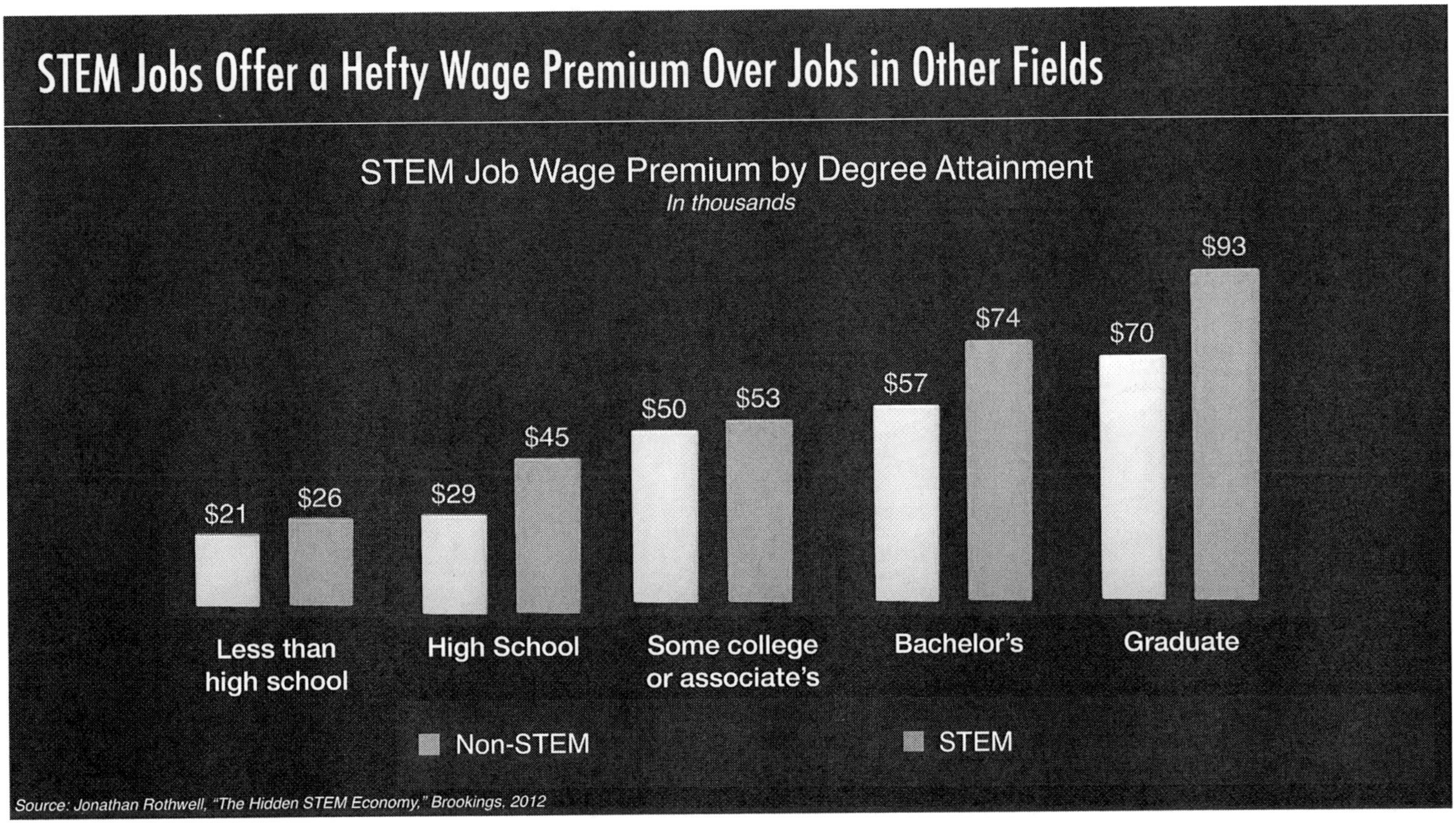
STEM Jobs Offer a Hefty Wage Premium Over Jobs in Other Fields
STEM Job Wage Premium by Degree Attainment
In thousands
$21
$26
$29
$45
$50
$53
$57
$74
$70
$93
Less than high school
High School
Some college or associate's
Bachelor's
Graduate
Non-STEM
STEM
Source: Jonathan Rothwell, "The Hidden STEM Economy," Brookings, 2012

All Goods Producing Industries—Not Just AI's—Still Matter

22%

share of economic output in goods producing industries

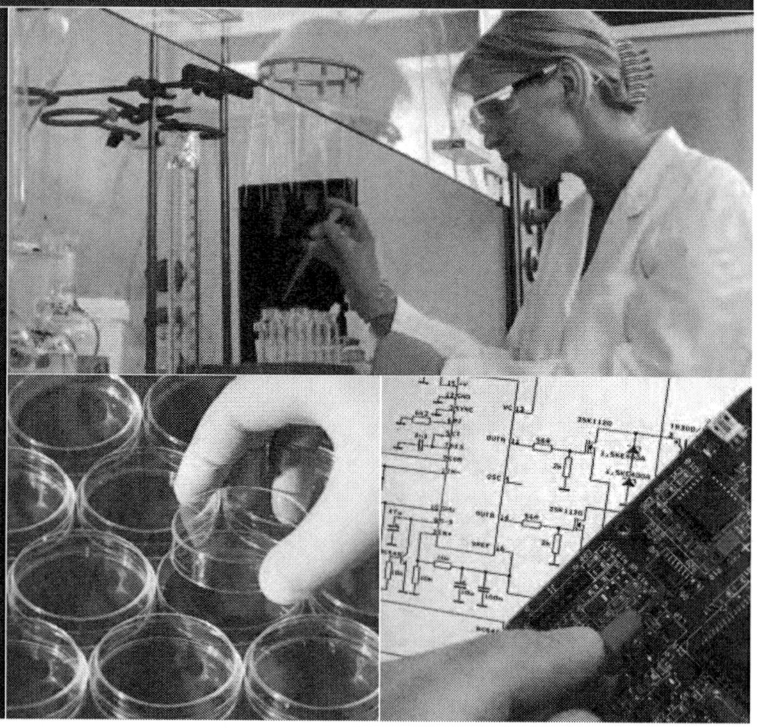

Source: Brookings Analysis of Moody's Economy.com Data

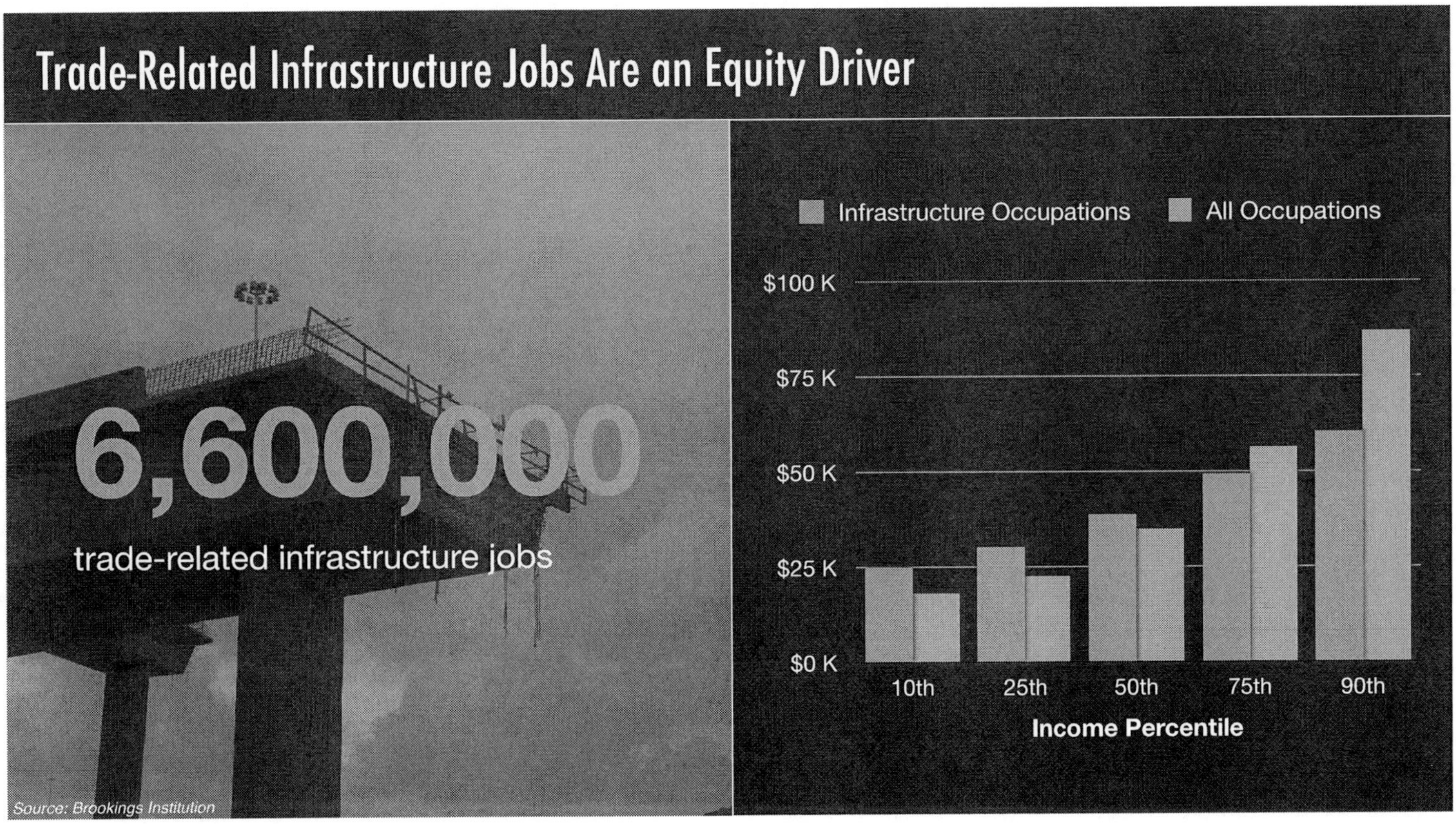

Trade-Related Infrastructure Jobs Are an Equity Driver
6,600,000
trade-related infrastructure jobs
Source: Brookings Institution
Infrastructure Occupations
All Occupations
$100 K
$75 K
$50 K
$25 K
$0 K
10th
25th
50th
75th
90th
Income Percentile

1 3

2 Trade prioritization requires a new perspective on freight trends

What Affects Trading Relationships?
Population
Complementary Industries
Logistics Workforce
Distance between Markets

Electronics trade demonstrates long-distance flows

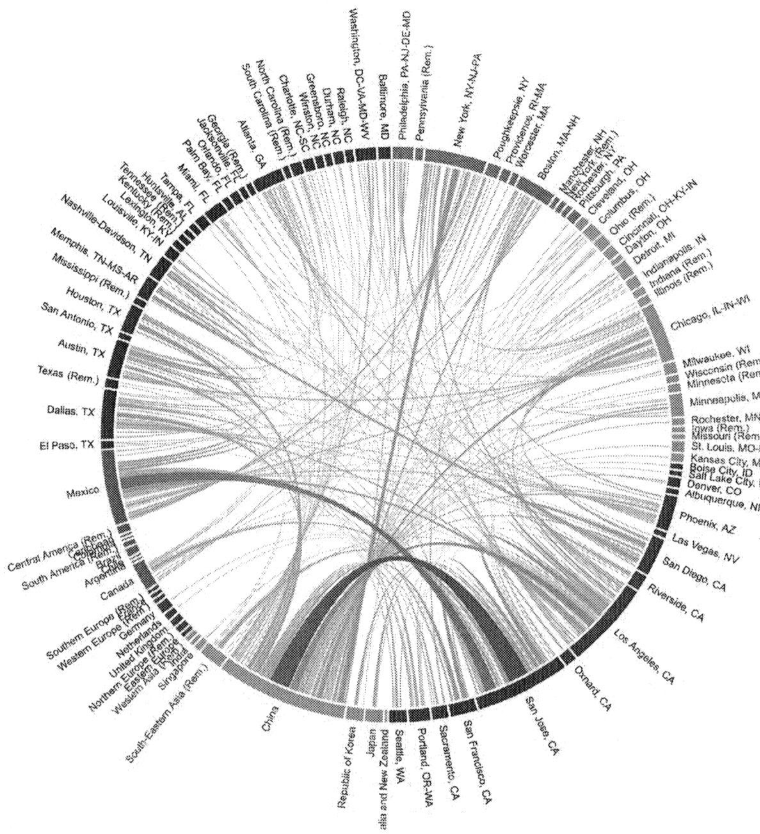

Source: Brookings Institution

Meanwhile, mixed freight travels shorter distances

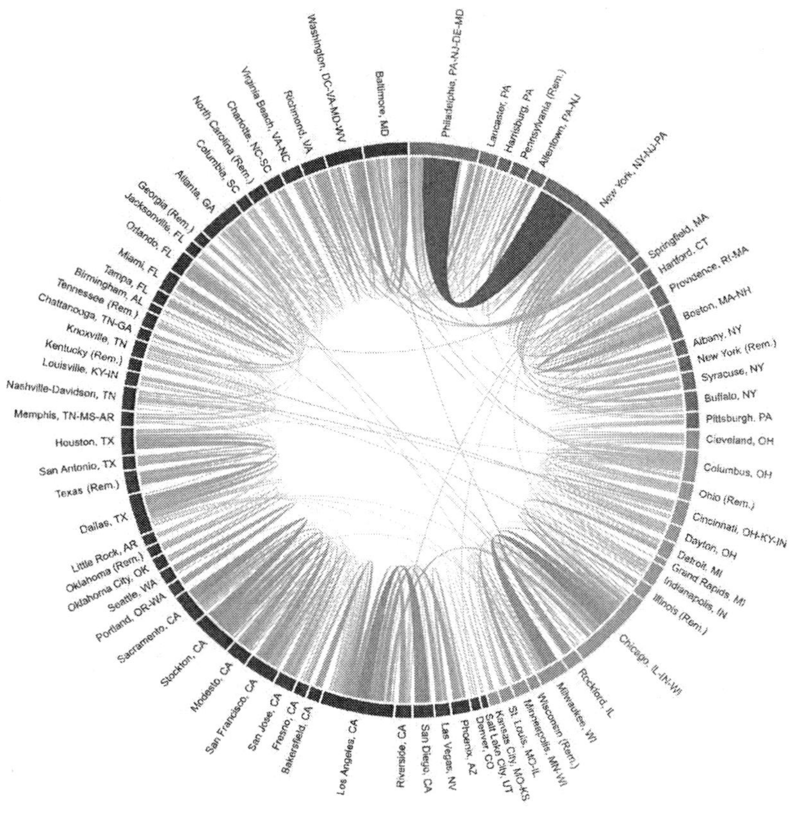

Source: Brookings Institution

77 percent

share of trade crossing state borders

Source: Brookings Institution

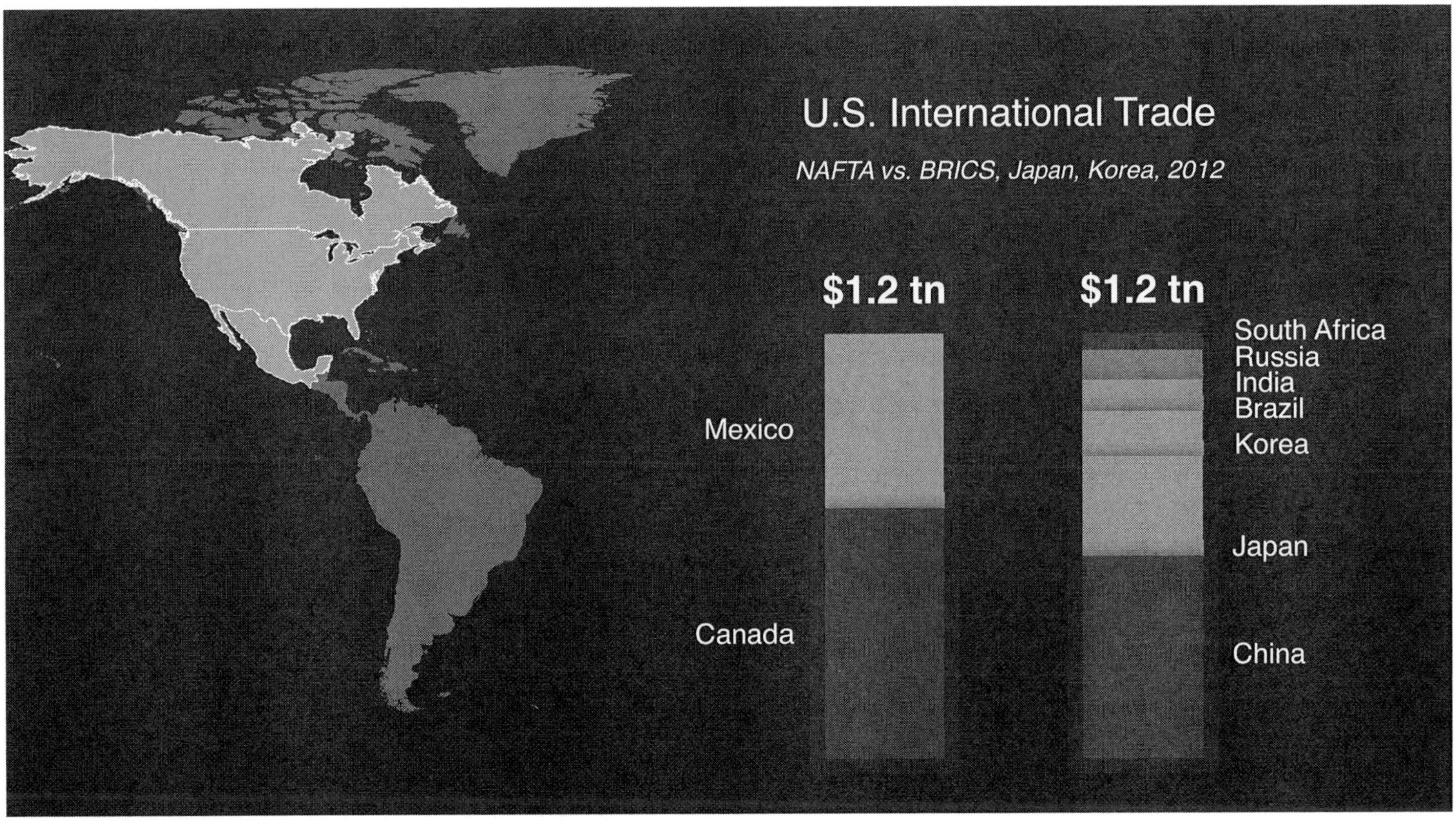
U.S. International Trade
NAFTA vs. BRICS, Japan, Korea, 2012
$1.2 tn
$1.2 tn
Mexico
Canada
South Africa
Russia
India
Brazil
Korea
Japan
China

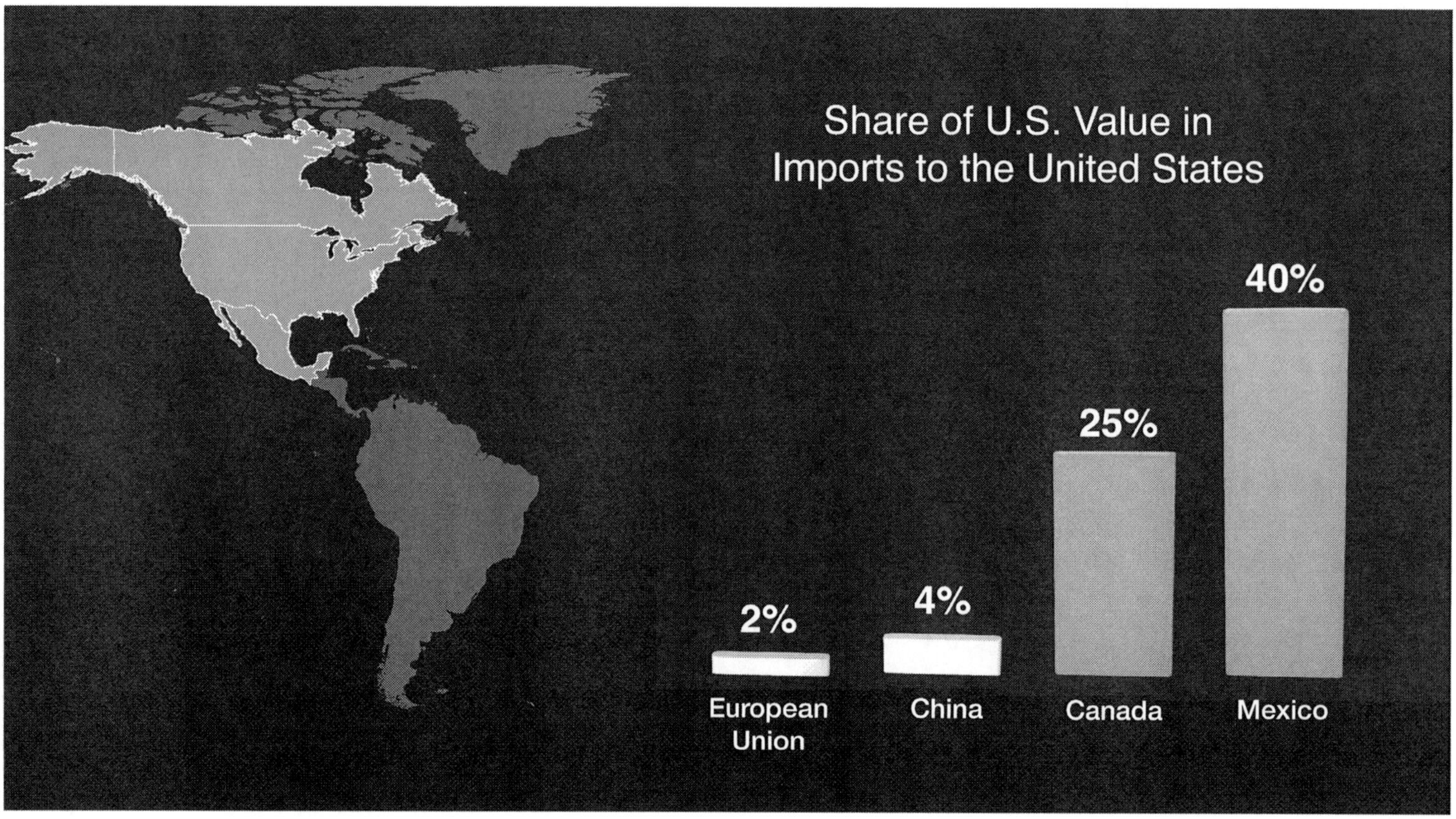
Share of U.S. Value in
Imports to the United States
2%
4%
25%
40%
European
Union
China
Canada
Mexico

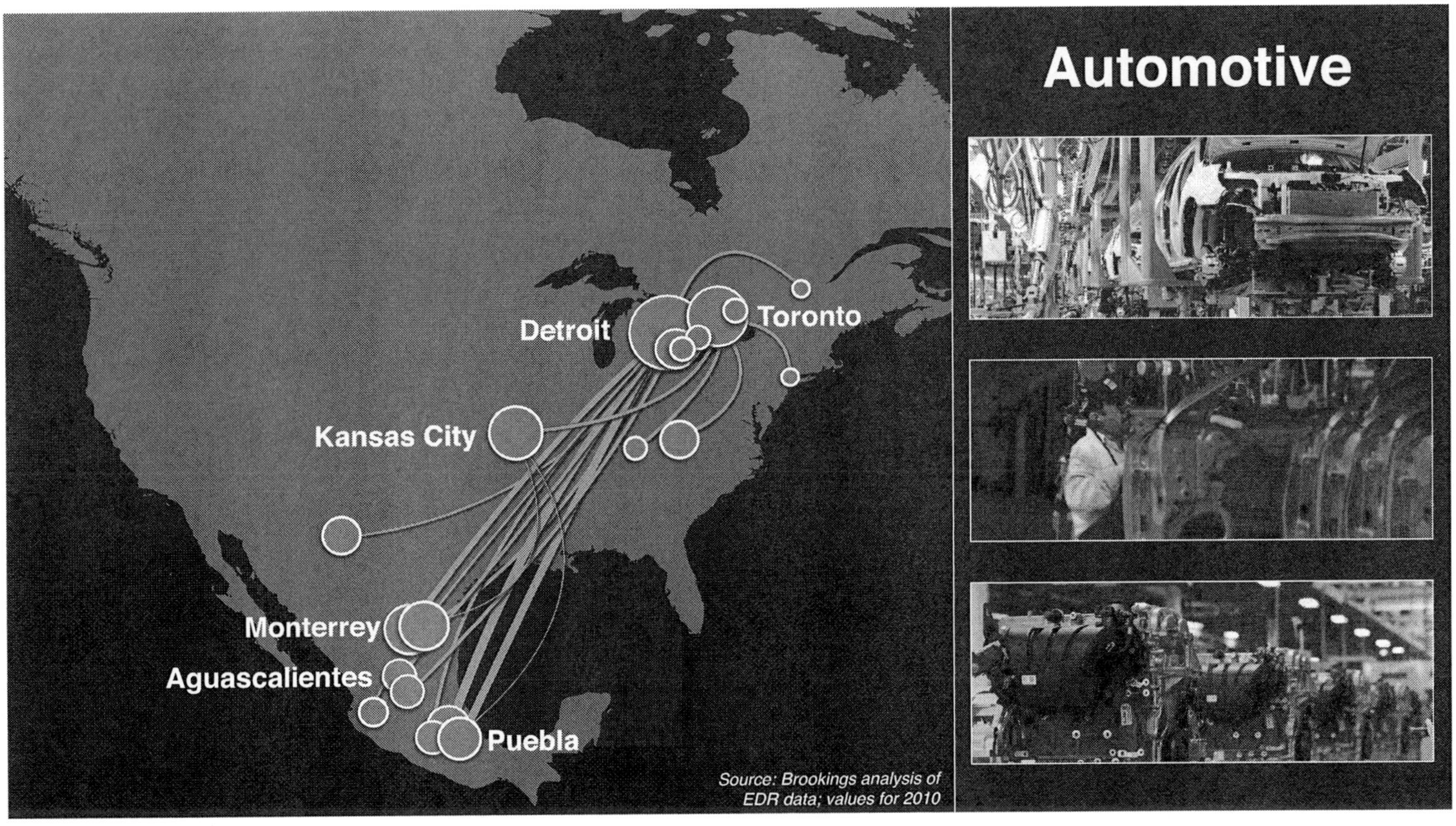
Automotive
Detroit
Toronto
Kansas City
Monterrey
Aguascalientes
Puebla
Source: Brookings analysis of EDR data; values for 2010

Detroit
$111 billion
Transportation Equipment

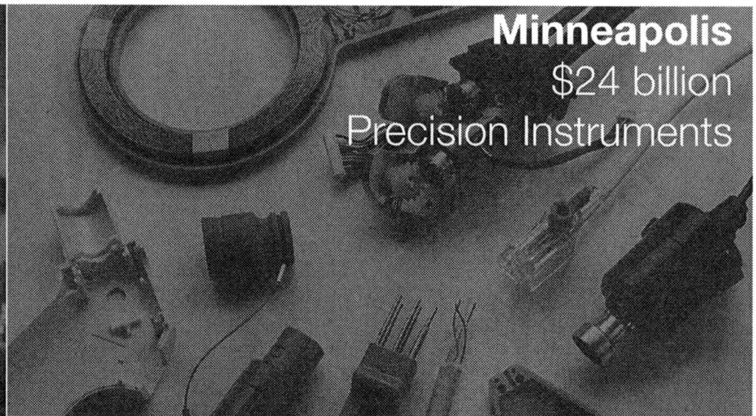
Minneapolis
$24 billion
Precision Instruments

San Jose
$102 billion
Electronics

Houston
$125 billion
Chemicals + Plastics

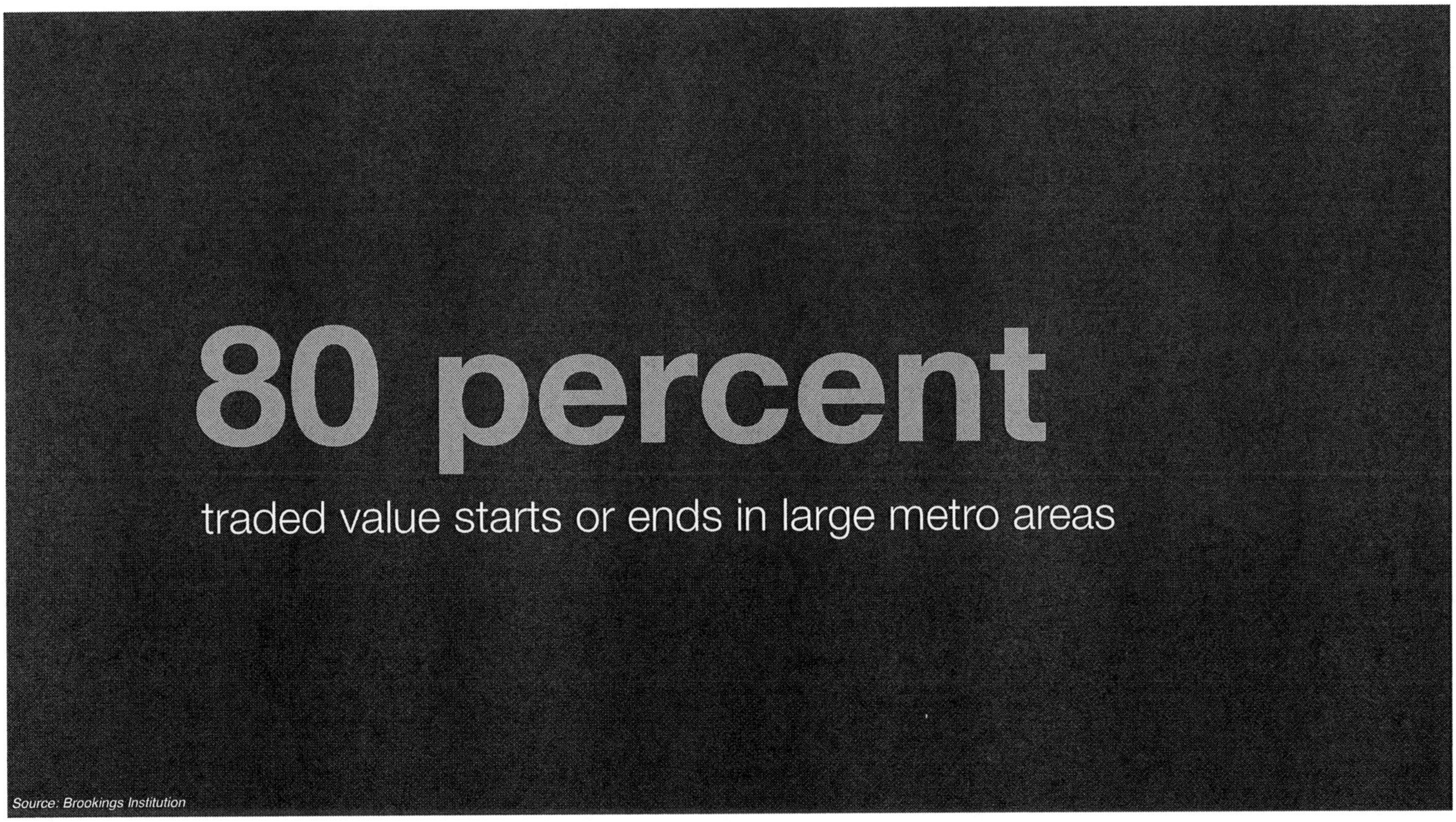
80 percent
traded value starts or ends in large metro areas
Source: Brookings Institution

Specific places matter in our freight network, many of which face growing congestion that ripples nationwide

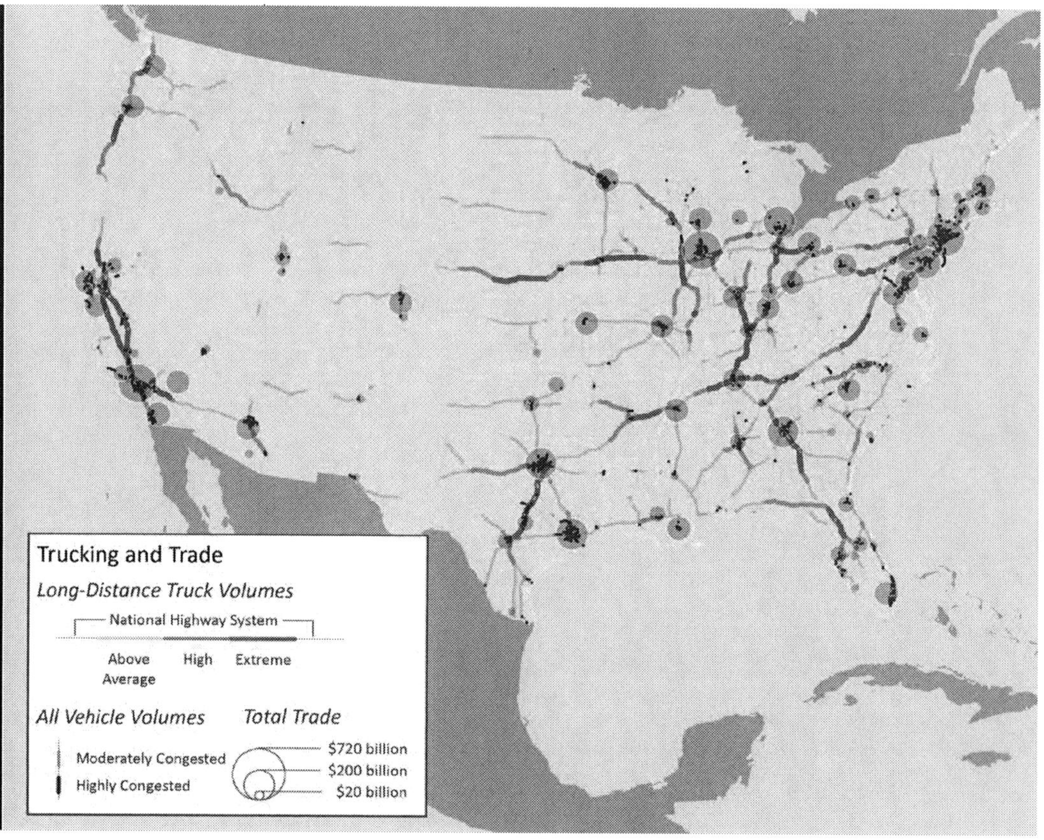

Source: Brookings MetroFreight Series and FAF

Congestion is especially bad in metro cores, limiting logistical efficiencies for key freight assets

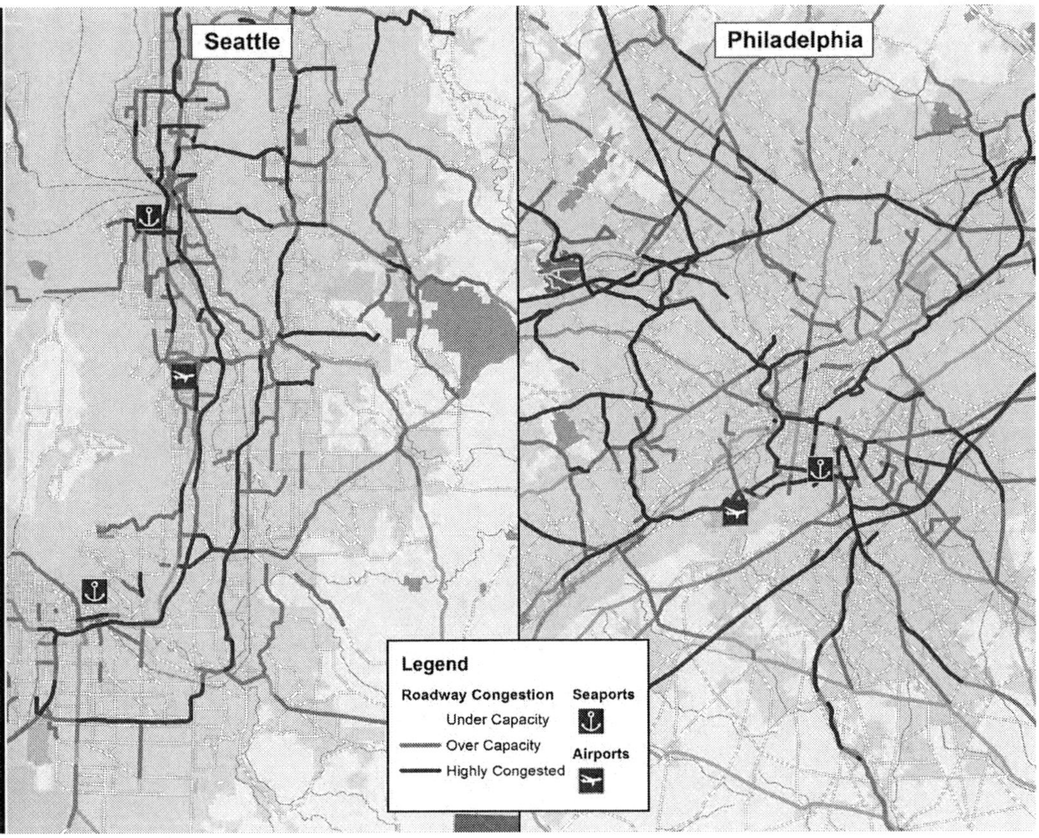

Source: Brookings MetroFreight Series and FAF

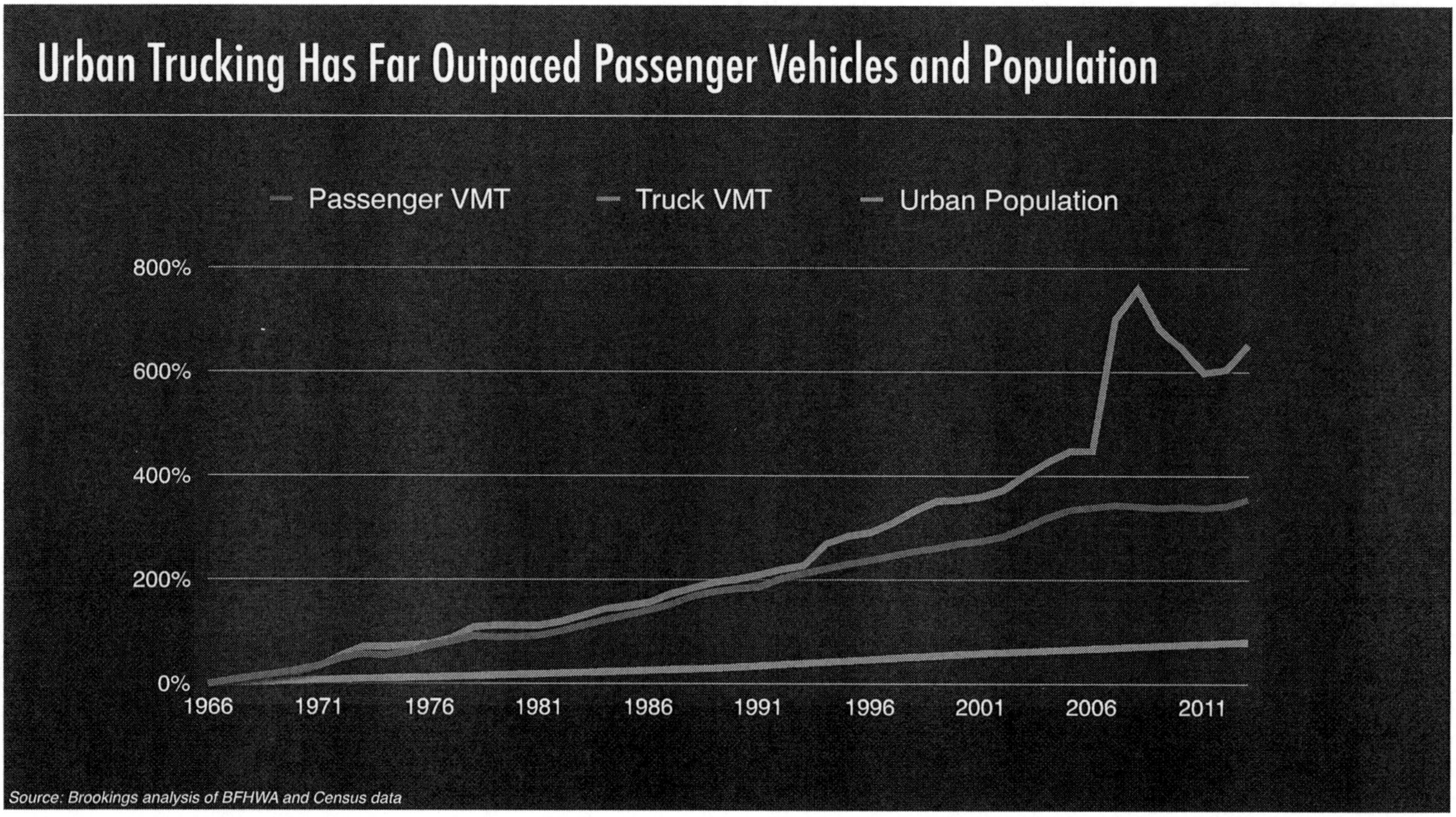
Urban Trucking Has Far Outpaced Passenger Vehicles and Population
Passenger VMT
Truck VMT
Urban Population
800%
600%
400%
200%
0%
1966
1971
1976
1981
1986
1991
1996
2001
2006
2011
Source: Brookings analysis of BFHWA and Census data

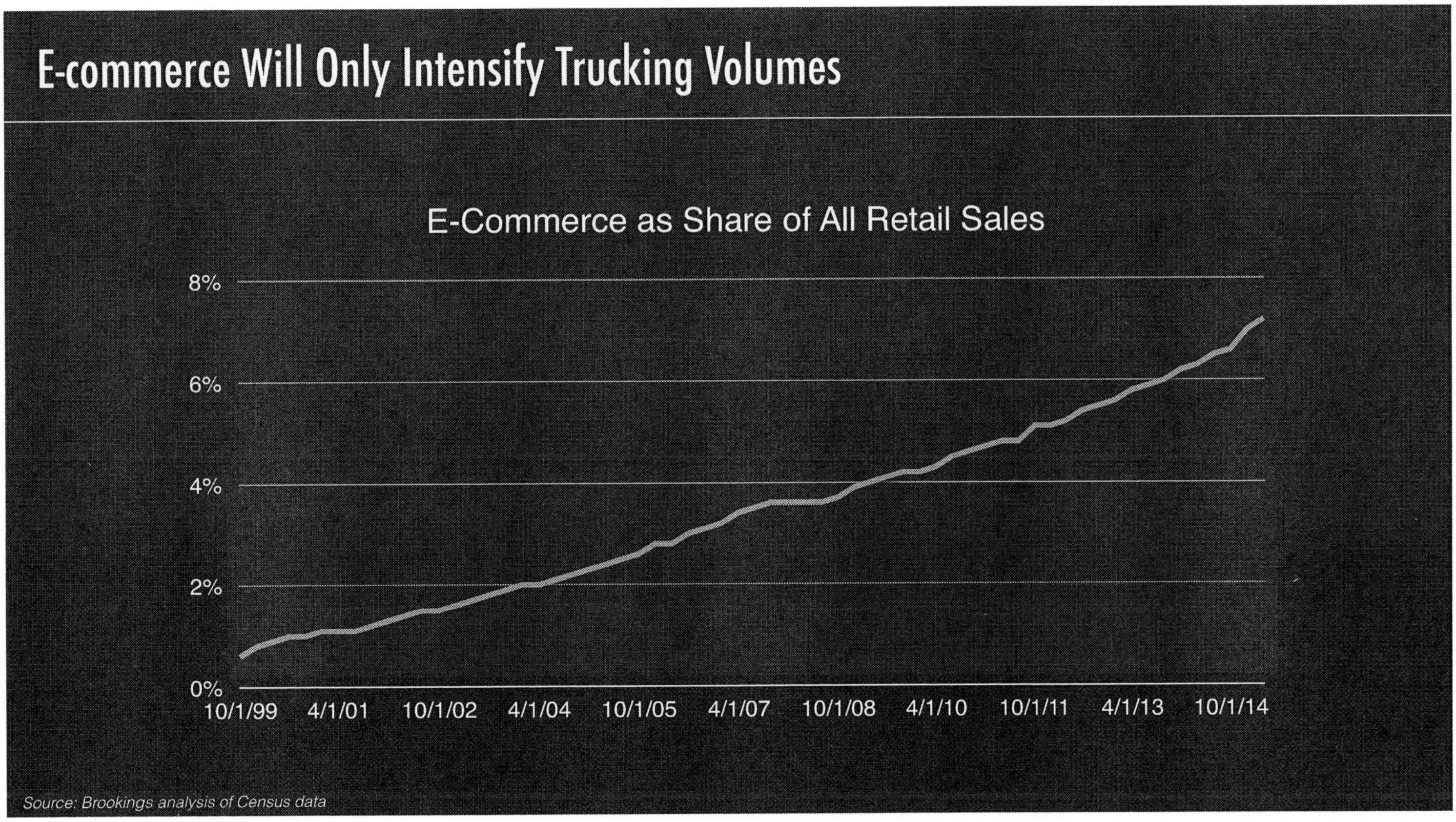
E-commerce Will Only Intensify Trucking Volumes
E-Commerce as Share of All Retail Sales
8%
6%
4%
2%
0%
10/1/99
4/1/01
10/1/02
4/1/04
10/1/05
4/1/07
10/1/08
4/1/10
10/1/11
4/1/13
10/1/14
Source: Brookings analysis of Census data

How does this play-out in Minnesota?

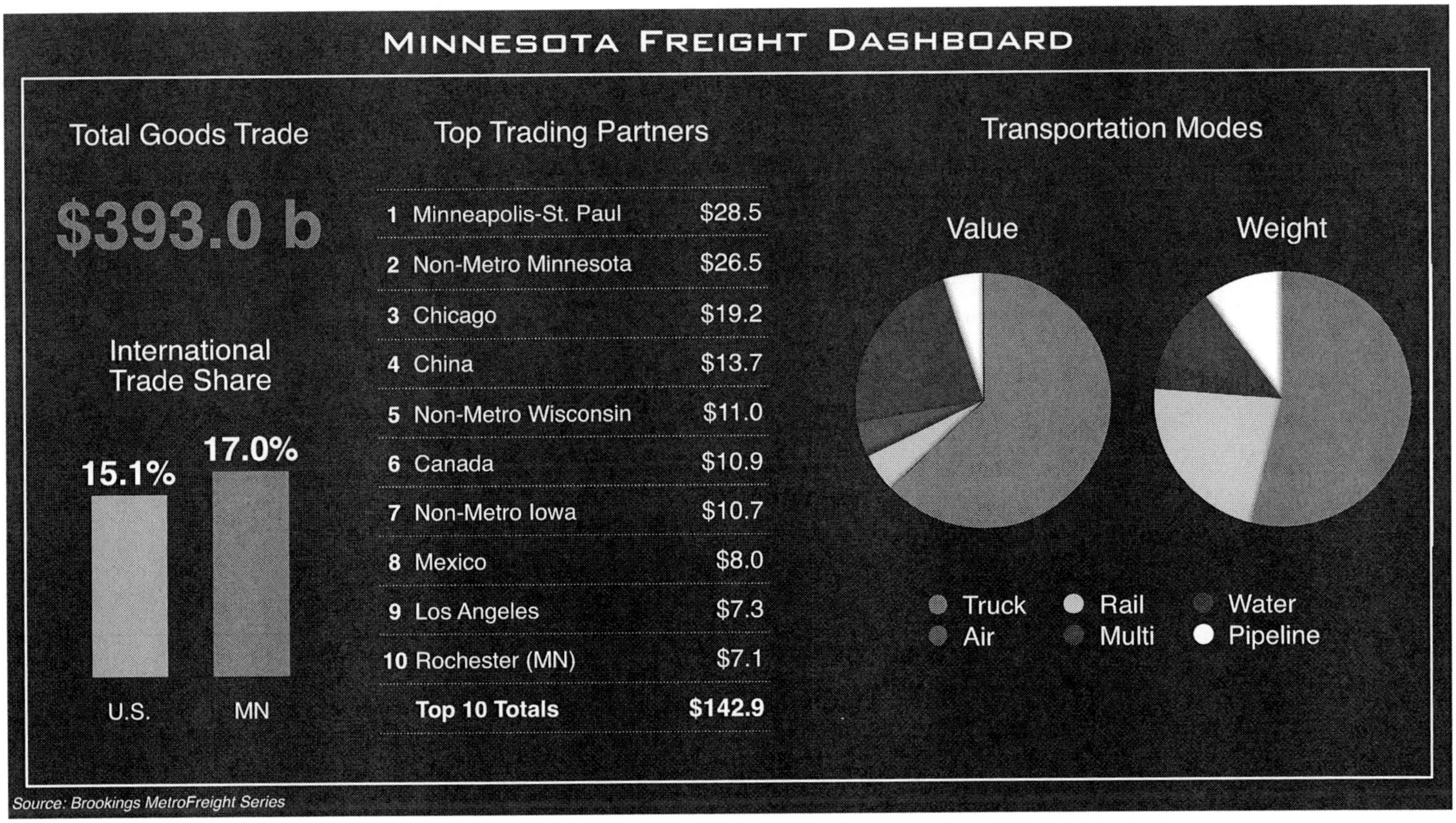
MINNESOTA FREIGHT DASHBOARD
Total Goods Trade
$393.0 b
International Trade Share
15.1%
17.0%
U.S.
MN
Top Trading Partners
1 Minneapolis-St. Paul $28.5
2 Non-Metro Minnesota $26.5
3 Chicago $19.2
4 China $13.7
5 Non-Metro Wisconsin $11.0
6 Canada $10.9
7 Non-Metro Iowa $10.7
8 Mexico $8.0
9 Los Angeles $7.3
10 Rochester (MN) $7.1
Top 10 Totals $142.9
Transportation Modes
Value
Weight
Truck
Rail
Water
Air
Multi
Pipeline
Source: Brookings MetroFreight Series

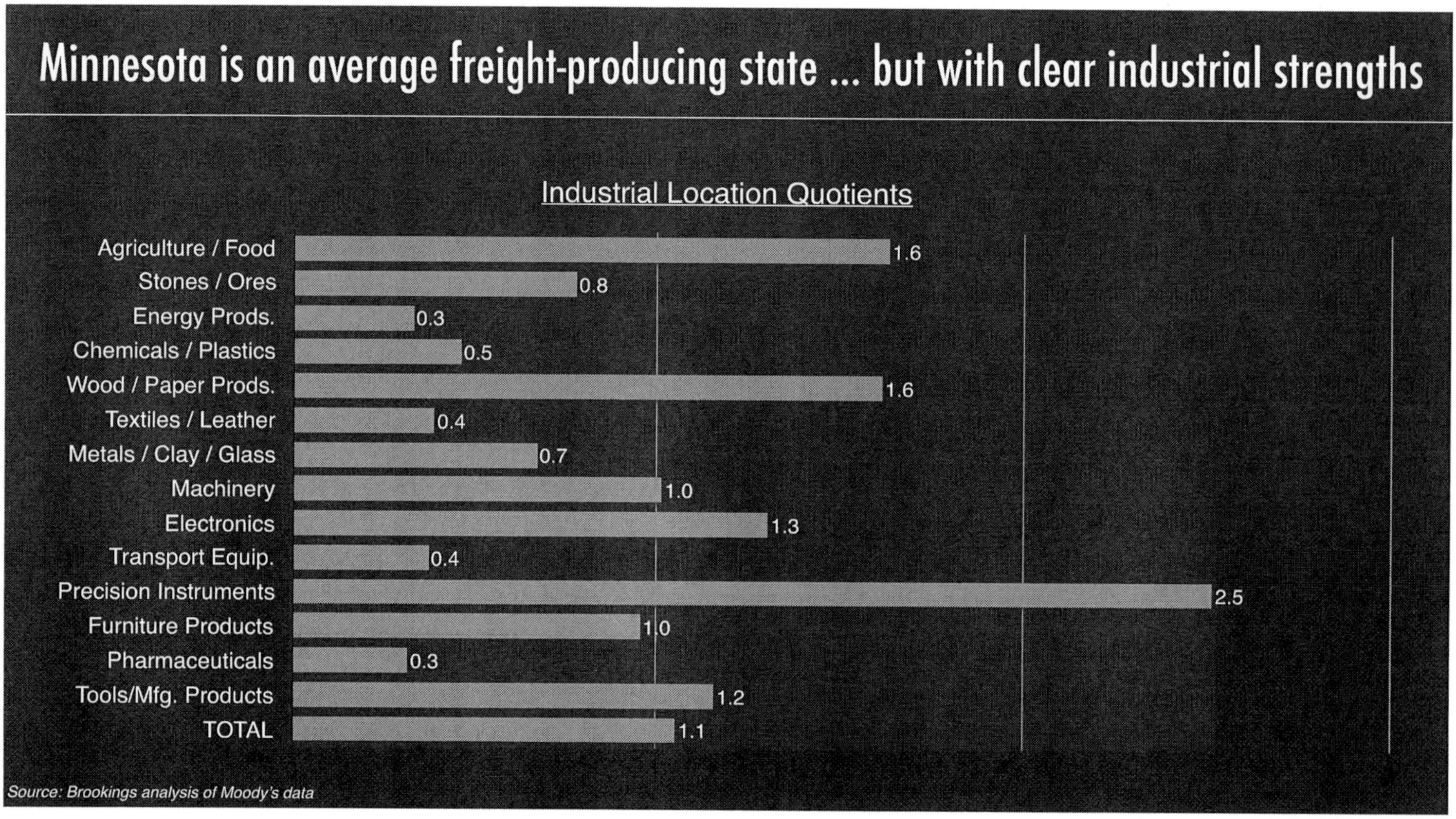
Minnesota is an average freight-producing state ... but with clear industrial strengths
Industrial Location Quotients
Agriculture / Food 1.6
Stones / Ores 0.8
Energy Prods. 0.3
Chemicals / Plastics 0.5
Wood / Paper Prods. 1.6
Textiles / Leather 0.4
Metals / Clay / Glass 0.7
Machinery 1.0
Electronics 1.3
Transport Equip. 0.4
Precision Instruments 2.5
Furniture Products 1.0
Pharmaceuticals 0.3
Tools/Mfg. Products 1.2
TOTAL 1.1
Source: Brookings analysis of Moody's data

Non-Metro Minnesota trades agriculture in surrounding region

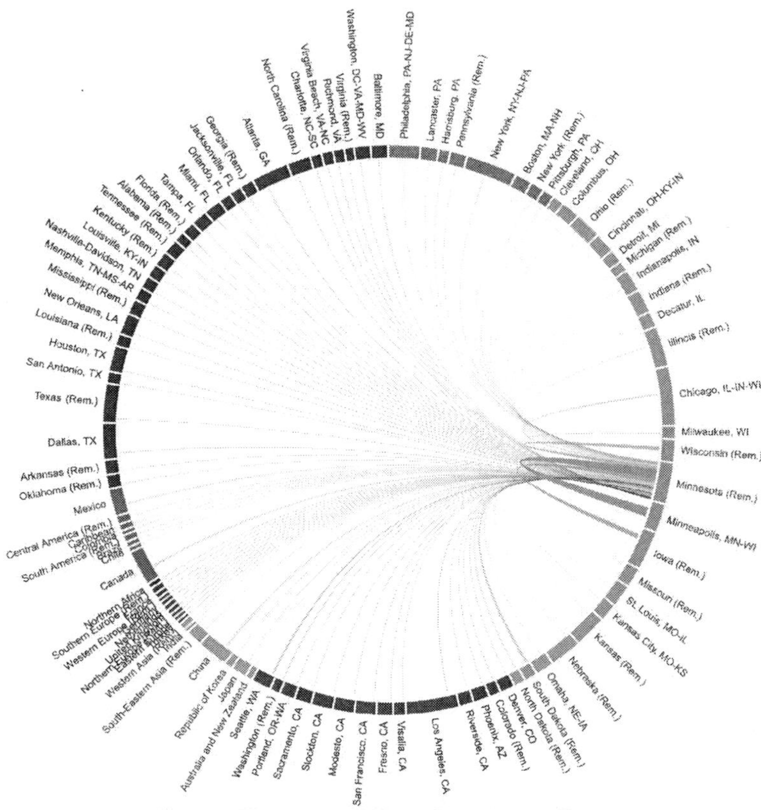

Source: Brookings Institution

Meanwhile, MSP's precision instruments are a more global industry

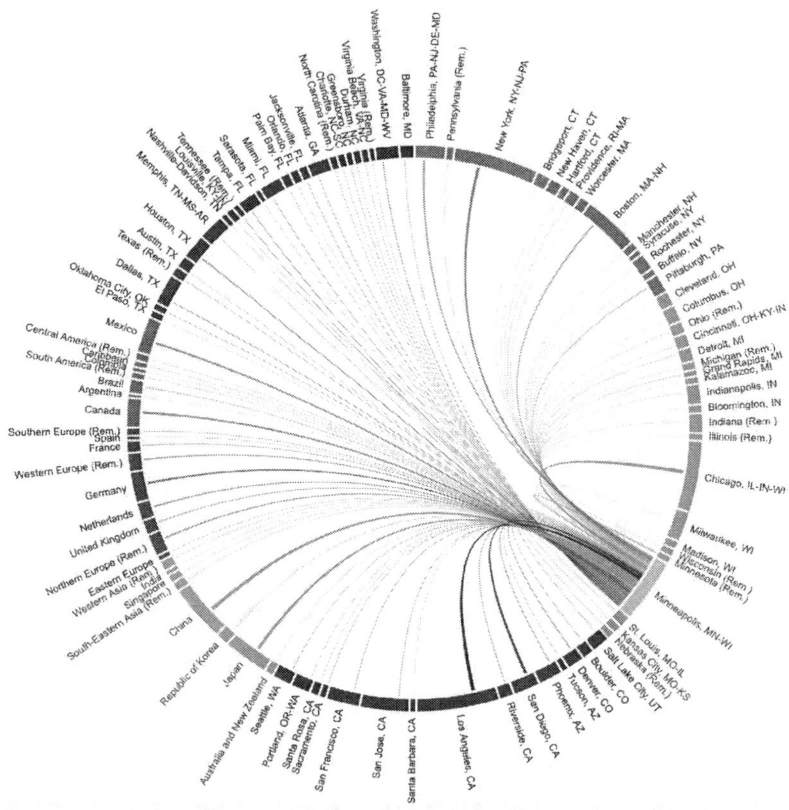

Source: Brookings Institution

1 2

3 What's the future of American freight planning and investment?

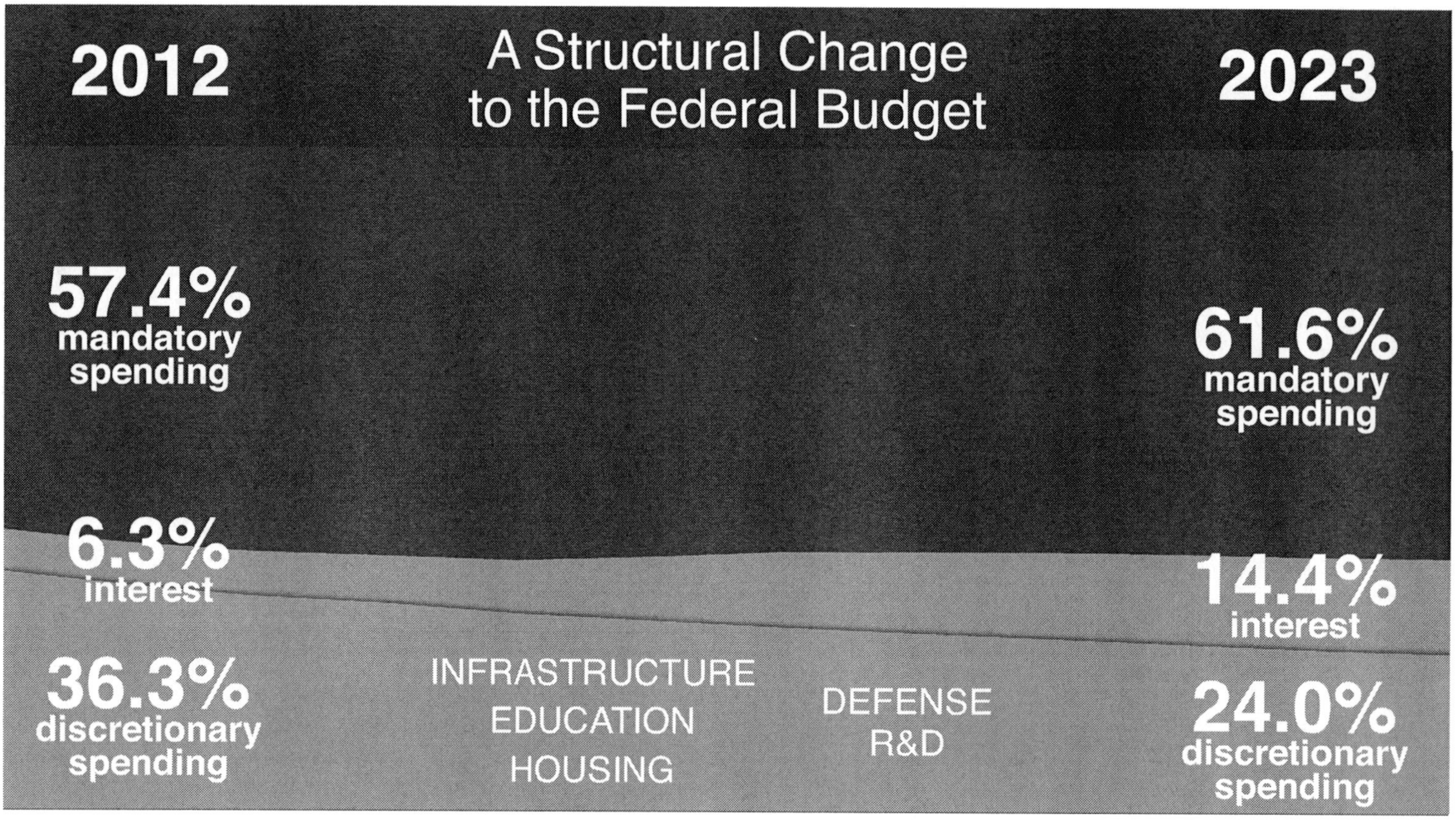
2012
A Structural Change
to the Federal Budget
2023
57.4%
mandatory
spending
61.6%
mandatory
spending
6.3%
interest
14.4%
interest
36.3%
discretionary
spending
INFRASTRUCTURE
EDUCATION
HOUSING
DEFENSE
R&D
24.0%
discretionary
spending

Federal and State Policies Sidestep Economic Priorities
Lack of Project Accountability
Ambiguous Metrics

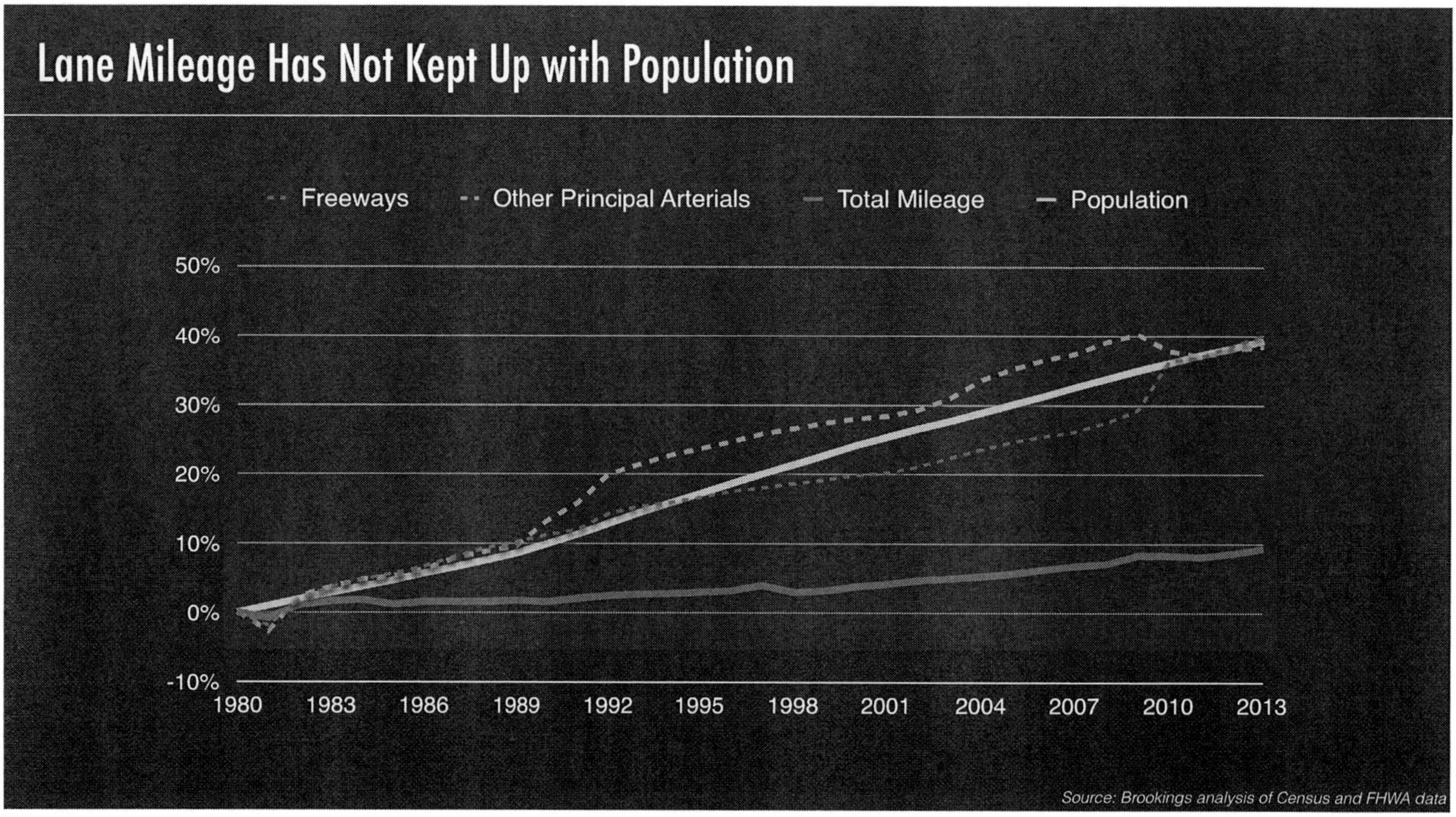
Lane Mileage Has Not Kept Up with Population
Freeways
Other Principal Arterials
Total Mileage
Population
50%
40%
30%
20%
10%
0%
-10%
1980
1983
1986
1989
1992
1995
1998
2001
2004
2007
2010
2013
Source: Brookings analysis of Census and FHWA data

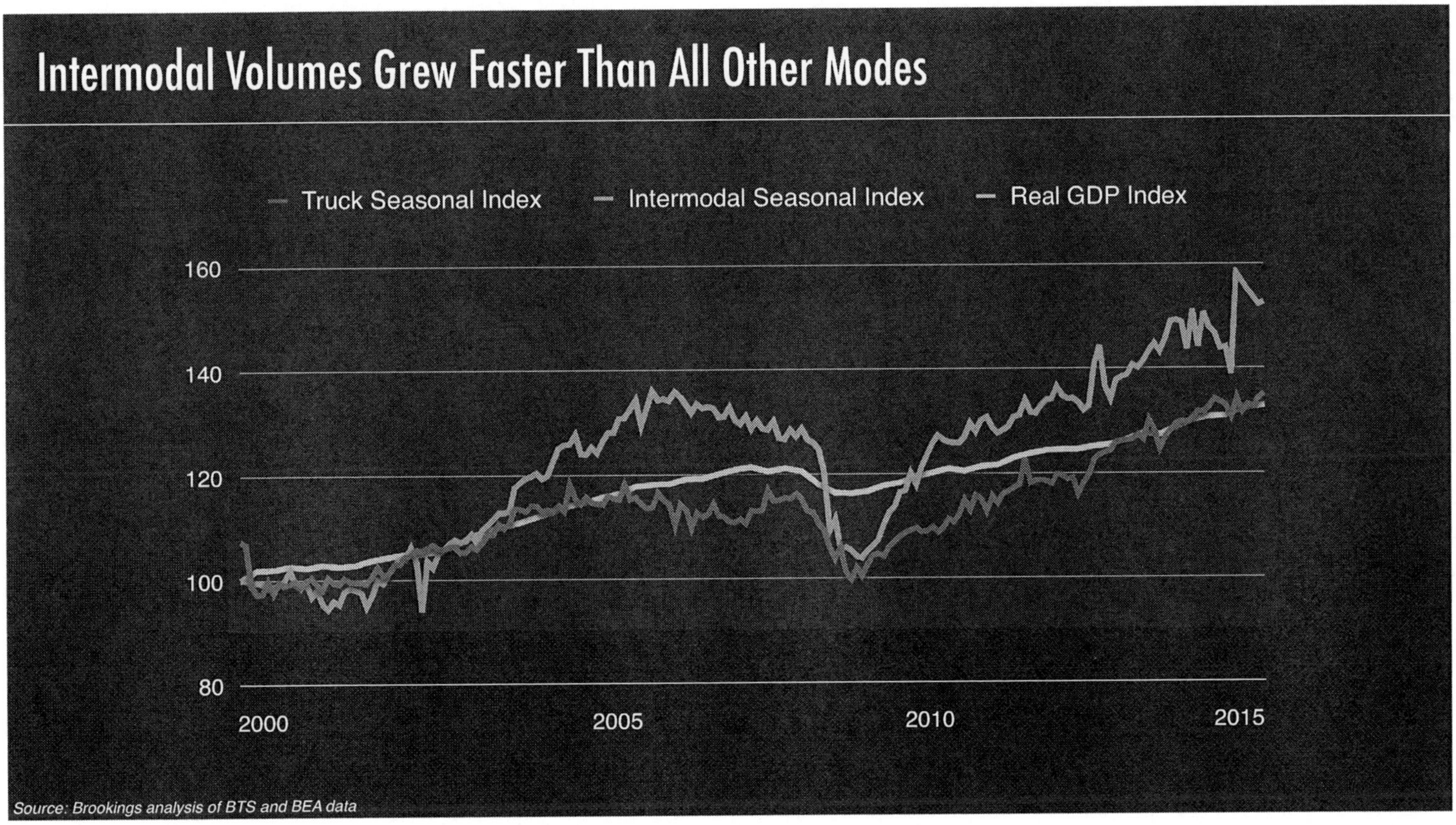
Intermodal Volumes Grew Faster Than All Other Modes
Truck Seasonal Index
Intermodal Seasonal Index
Real GDP Index
160
140
120
100
80
2000
2005
2010
2015
Source: Brookings analysis of BTS and BEA data

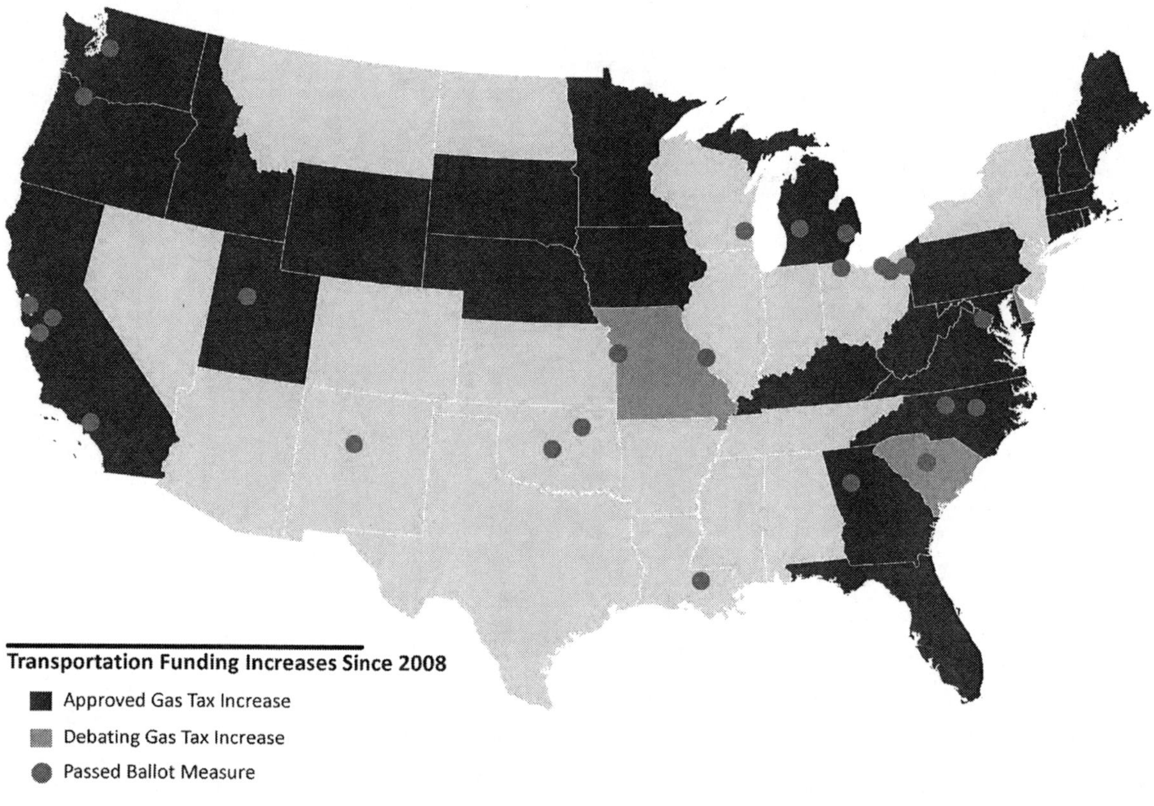
Transportation Funding Increases Since 2008
Approved Gas Tax Increase
Debating Gas Tax Increase
Passed Ballot Measure

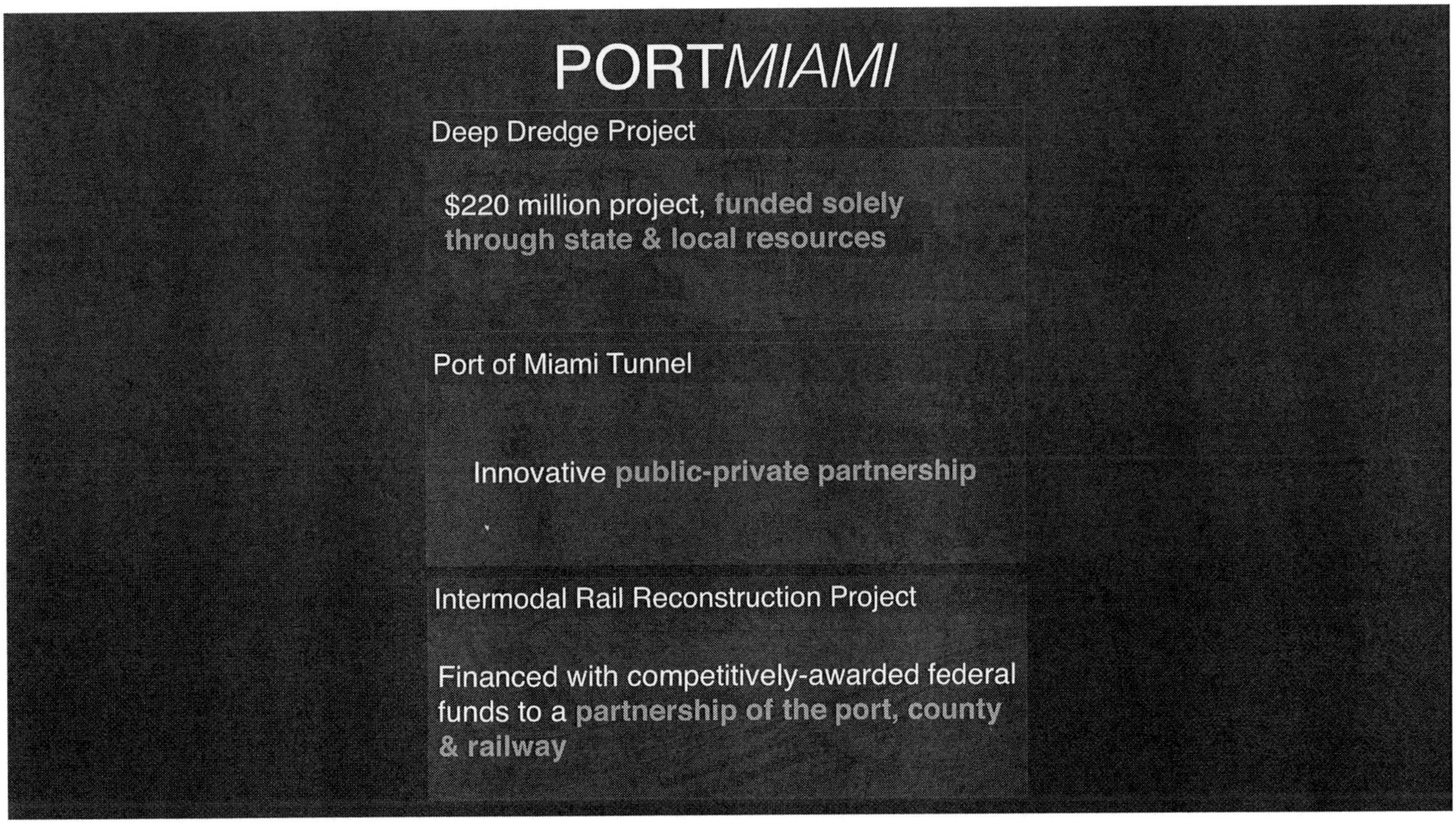
PORT*MIAMI*
Deep Dredge Project
$220 million project, funded solely through state & local resources
Port of Miami Tunnel
Innovative public-private partnership
Intermodal Rail Reconstruction Project
Financed with competitively-awarded federal funds to a partnership of the port, county & railway

How do we get more done?

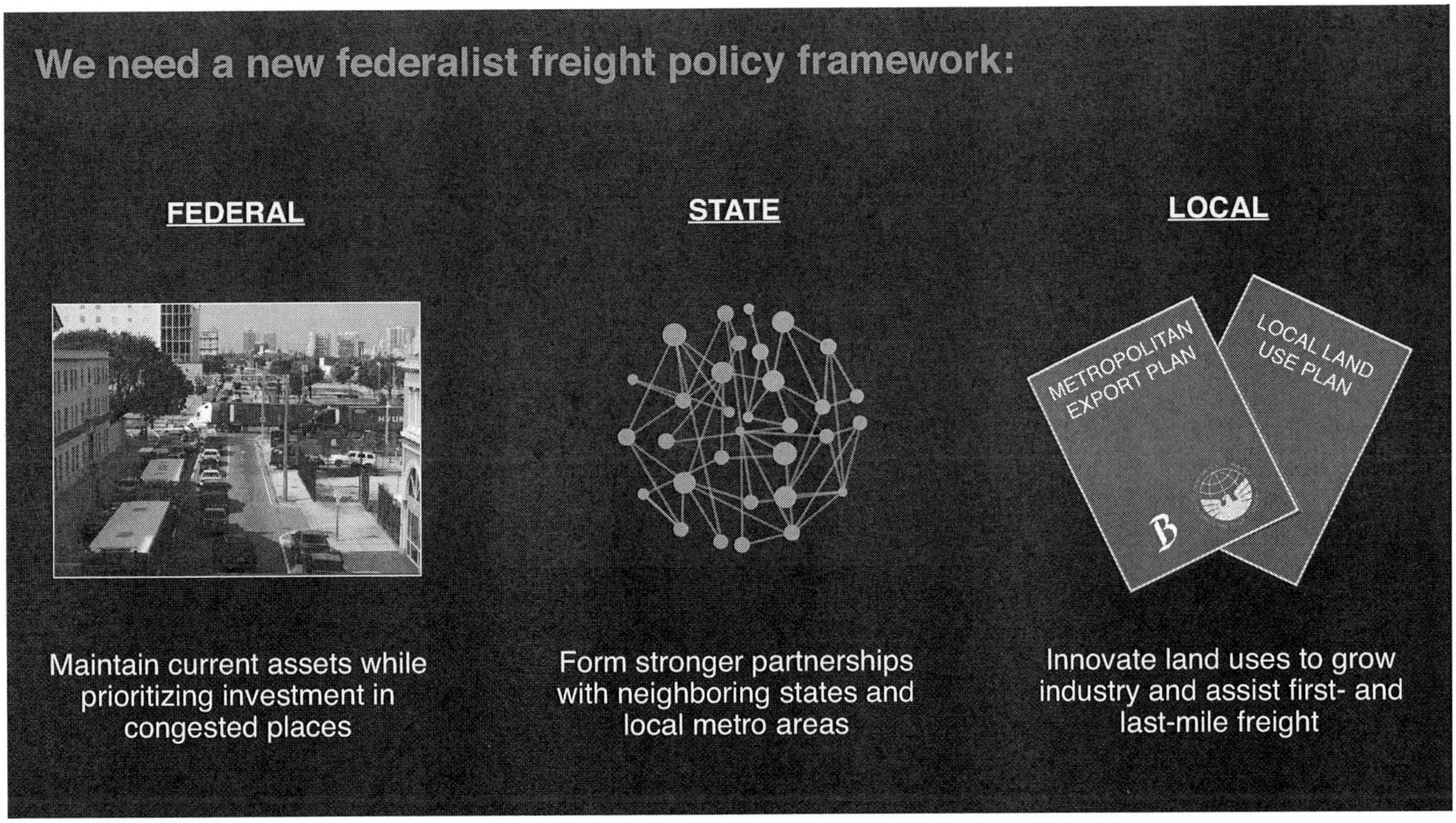
We need a new federalist freight policy framework:
FEDERAL
STATE
LOCAL
METROPOLITAN EXPORT PLAN
LOCAL LAND USE PLAN
Maintain current assets while prioritizing investment in congested places
Form stronger partnerships with neighboring states and local metro areas
Innovate land uses to grow industry and assist first- and last-mile freight

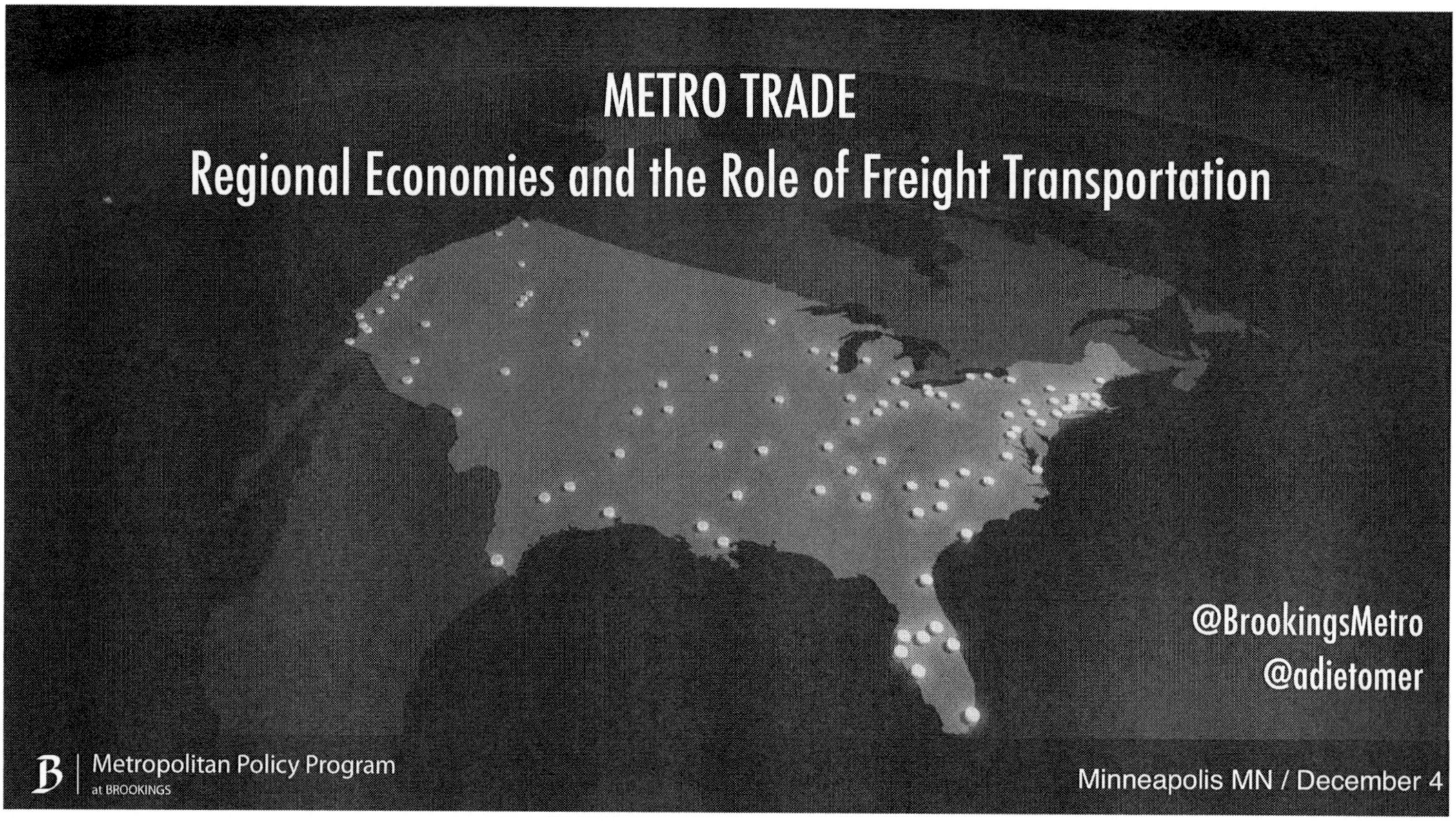
METRO TRADE
Regional Economies and the Role of Freight Transportation
@BrookingsMetro
@adietomer
Metropolitan Policy Program
at BROOKINGS
Minneapolis MN / December 4